The Man in the White Sharkskin Suit

ALSO BY LUCETTE LAGNADO

*Children of the Flames: Dr. Josef Mengele and
the Untold Story of the Twins of Auschwitz*

Edith and Leon, Cairo 1943.

THE MAN IN THE WHITE SHARKSKIN SUIT

My Family's Exodus
from Old Cairo to the New World

· · · · ·

LUCETTE LAGNADO

An Imprint of HarperCollinsPublishers

HarperCollins books may be purchased for educational, business, or sales promotional use. For information, please write: Special Markets Department, HarperCollins Publishers, 10 East 53rd Street, New York, NY 10022.

Grateful acknowledgment is made for permission to reproduce the illustrations on the following pages: Pages iv, 9, 22, 38, 90, 108, 142, 229, 230, and 245: courtesy of the Lagnado family archives. Pages 15, 55, 63, 164, 166, 170, 194, 232, 235, 271, and 314: courtesy of César Lagnado. Pages 41 and 51: courtesy of Salomone Silvera. Page 71: courtesy of Pico and Rachel Hakim. Page 77: courtesy of David Ades. Page 105: courtesy of Edouard Lagnado. Page 131: courtesy of *Community Times*, September 2006, "GROPPI: People's Memories of the World's Ritziest Tea-Room." Page 287: courtesy of Anne and Burton Lee. Page 325: courtesy of the author.

FIRST EDITION

Designed by Mia Risberg

Library of Congress Cataloging-in-Publication Data is available upon request.

ISBN: 978-0-06-082212-5
ISBN-10: 0-06-082212-0

07 08 09 10 11   DIX/RRD   10 9 8 7 6 5 4

To my husband, Douglas Feiden,
and to the memory of Leon and Edith

And the Children of Israel wept and said: "Who will feed us meat? We remember the fish that we ate in Egypt free of charge, and the cucumbers, melons, leeks, onions, and garlic. But now, our life is parched, and there is nothing. We have nothing to anticipate but manna."

—Numbers 11:4–6

It was then that I stood up in the theater and shouted: "Don't do it. It's not too late to change your minds, both of you. Nothing good will come of it, only remorse, hatred, scandal, and two children whose characters are monstrous."

—Delmore Schwartz, *In Dreams Begin Responsibilities*

CONTENTS

A Courtship in Cairo—

SPRING 1943

Edith was seated outdoors at La Parisiana, Cairo's most popular café, enjoying a *café turque* with her mother, when she noticed the man in white. He was looking her way and smiling, and though he, too, was sitting down, she could tell he was extremely tall. He raised his glass and tipped it in her direction. She was so shy, she quickly turned her head, not daring to return his glance.

There was never room for the mildest flirtation in Edith's life. Her mother, Alexandra, was always by her side, policing her every move, so strict she didn't permit her daughter to have any dealings with men that even hinted at romance. At twenty, Edith had never had a suitor. She wasn't allowed to engage in the light, friendly banter that was encouraged between the sexes in wartime Cairo, a culture that managed to be both old-fashioned and libertine at the same time.

Early on, her mother had laid down the law.

Edith was expected to come straight home from work at the end of the day. She couldn't socialize with colleagues, especially male colleagues, and she was to shun any and all advances from the eminently respectable bachelors who taught alongside her at the École Cattaoui. A teacher of children, Edith was treated like a child in her own home.

She was so meek, she never chafed at the restrictions. She was simply grateful to have a job at the distinguished private school that had hired her when she was barely fifteen, and whose main benefactors were Moise Cattaoui, a Jewish Pasha, and one of the wealthiest men in Egypt, along with his socially prominent wife, Madame Cattaoui Pasha, who was the queen's lady-in-waiting.

Of course, Alexandra had never reckoned with anyone like the man in white, and neither had her woefully naive daughter.

At forty-two, Leon was used to getting his way, especially with women. He had never been married and, like Edith, he lived at home with his mother. But the resemblance ended there. Unlike her, he suffered from no restrictions on his life whatsoever.

Cairo had a million diversions, and Leon took advantage of every one of them. He relished being single, venturing out every night and not returning until dawn. He ambled elegantly through the city in constant search of entertainment. Dining, dancing, and gambling were his great loves, and he wandered from restaurants to cafés to dance halls to casinos. It was 1943, the height of World War II, and the streets and the cinemas and the nightspots were crowded with British soldiers in their khaki uniforms and jaunty berets, which suited Leon fine because he didn't love anyone as much as he loved *les Anglais*.

Wherever he went, he stood out, a towering figure in expensive, hand-tailored suits made of white sharkskin.

The soft shiny material was all the rage among Cairo's privileged classes.

He'd stop and catch his breath only on Friday night, the Jewish Sabbath, because he took religion as seriously as his games and pastimes. Early on, Leon had figured out a way for these seemingly contradictory sides of his nature, his love of God and his passion for pleasure, to coexist. He was a regular at temple on Friday night and Saturday morning, but come Saturday night, his exuberant, frivolous life resumed and continued uninterrupted throughout the week.

In contrast, most of Edith's evenings were spent quietly at home in Sakakini, a poor section of Cairo, with her mother and her younger brother, Félix, as her sole companions. If she wanted to see a movie or go to a café, it was arm in arm with Alexandra. Dance halls, cabarets, nightclubs, were off-limits. There were no escapes and few pleasures for the young woman, except for the books she devoured.

She worked so hard at the École Cattaoui, she caught the eye of its famous patron. Madame Cattaoui Pasha, intrigued by her diligent, lovely recruit, offered her a job as librarian of the Bibliothèque Cattaoui. It was an extraordinary opportunity. The pasha's wife had a vision she wanted Mademoiselle Edith to realize: to build a school library that would house all the great French classics.

Still a teenager, operating on instinct since she had no formal training as a librarian, Edith went on a buying spree, purchasing hundreds of books. After months of feverish acquisitions—Flaubert, Proust, Balzac, Zola—she was able to report that the collection was almost complete.

Madame Cattaoui Pasha was so pleased, she gave the young woman a gift: a key to the library—enormous, brass, shiny, ornate. Edith's hand trembled as she accepted it. It was the single greatest honor she would ever receive. For Edith, it was as if she had been handed the Keys to the Kingdom.

Leon had no patience for the contemplative life, and the only books he pored through were his prayer books and his Bible, though his favorite reading material was probably *La Bourse Egyptienne,* the popular financial newspaper that tracked the Egyptian stock market.

He began each day by praying with fellow Jews. He did business with French Colonial merchants and Greek entrepreneurs. He gambled with wealthy Egyptians, including, on occasion, the king. And he socialized with the British officers stationed throughout Cairo. Always stylish and meticulously dressed, with an easy manner and a fluent command of English, Leon was one of the few outsiders they welcomed into their fold.

They even had an affectionate nickname for him: "Captain Phillips." No one knew its origins, but it was quintessentially British, and it stuck because he wore it so well. All around Cairo Leon became known as "The Captain." The French called him "Le Capitaine."

Cairo came alive at night. The workday ended early because it was so hot in the afternoon. People returned home from the office, took a long nap, and woke up refreshed and energized enough to go out again. The picture shows at the dozens of outdoor cinema houses didn't even start until nine. It was not uncommon to have dinner at eleven. No self-respecting belly dancer would even think of making an appearance until the stroke of midnight.

Throughout the city, popular eateries and dance halls set aside at least one table for King Farouk. It was off-limits to everyone else, with a discreet "reserved" sign, in case the portly young monarch decided to drop by. Farouk loved the nightlife and the women that came with it.

Leon and the king had that in common—along with a deep and abiding passion for poker.

One night, at a casino, Leon was invited to sit at the king's table and join a poker game under way.

On one round, Leon had a flush to Farouk's three kings. He reached for the pot, but the king grabbed his arm and stopped him.

He had four kings, Farouk declared, which beats a flush.

Leon frowned. He looked again at the cards in the monarch's hands, but he still saw only three kings.

Farouk burst out laughing and grabbed the chips at the center of the table like the greedy child he was.

"C'est moi le quatrième roi!" he exclaimed; Me—I am the fourth king!

Everyone at the table laughed along with him: The king was notoriously fond of such antics; he had played this card trick before.

Later, Leon would airily downplay the amount of money he had lost as bagatelle—small change. It had been a delightful evening, and the king had given him priceless material he could use to regale his British friends. They tended to cast a jaundiced eye on young Farouk, whose corruption in every sphere of life, even friendly poker games, was the stuff of legend.

But on this balmy spring afternoon at La Parisiana, Leon eyed Edith carefully, observing how she looked and dressed and carried herself. He liked to take his time and study a woman before making his move. Like Cary Grant, his idol, the actor to whom he bore more than a passing resemblance, he had high standards and definitive tastes. Only brunettes, and only brunette beauties, would do. Leon was even choosy about the film stars he watched. He could never be dragged, for instance, to a Katharine Hepburn movie—he found her so unappealing that he literally refused to see her movies. Edith brought to mind those *grandes beautés*—Vivien Leigh, Hedy Lamarr—but her kindred soul, her twin, was Ava Gardner, who would soon dominate Cairo's cinema houses.

Now, La Gardner, that was different: "C'est une grande beauté," he'd say.

Like the movie star, Edith had black wavy hair, a striking chiseled face, doleful eyes, and a regal bearing. She sat ramrod straight at the café table, her legs crossed, taking small, dainty sips of black coffee. She managed to be exotic without being overpowering. She was also deliciously young.

With a wave of his hand, Leon summoned the maître d' and gave him a note he'd dashed off on the back of the check. He instructed the waiter to slip it to that young woman over there—the pretty one sitting with her mother—and stand by her side as she read it. Leon tipped him generously.

The maître d' walked over to Edith's table, delicately placed the folded note in her hand, and bowed.

It consisted of only two lines:

"I find you very beautiful. Would it be possible for us to meet?"

Edith looked up and finally met his gaze. The man in white raised his drink her way once again. Though white jackets weren't uncommon in 1940s Cairo, he alone among the diners was wearing an entire suit of white. At last, she smiled.

This is how Leon and Edith's courtship began. It was like a scene from a movie with a perfect ensemble cast: a beautiful, shy ingénue, her overly protective mother, an affable and energetic restaurant maître d', and an aging roué who suddenly fancied himself in love.

Edith immediately passed the folded piece of paper to her mother, who frowned as she peered across the restaurant. Alexandra was only slightly older than Leon and passionately devoted to her daughter. Pretty, bookish Edith was a delicate flower that must be shielded at all cost.

If Alexandra let down her guard for a moment, if she didn't impose strict rules, legions of men were sure to hurt and betray the young woman.

The way that Alexandra had been hurt and betrayed by Edith's father.

Years earlier, he had left her to fend for herself in a culture noticeably hostile to women alone. There were few recourses for an abandoned wife—no alimony or child support payments. There she was, a mother of two small children with no resources. The family had nearly starved. Alexandra could survive perfectly well on cigarettes and coffee, but young Edith and her brother, Félix, were constantly hungry.

It was a miracle that Edith managed to finish school and snare a coveted teaching post at the École Cattaoui. For years, her salary supported all of them. Alexandra had never worked, and from an early age, Félix seemed unable to maintain a steady job and more interested in the fast bucks that came from petty cons. Because of Edith's diligence, the family now had more to eat than they'd had in years, and occasional indulgences, like afternoons at La Parisiana.

All that, and more, passed through Alexandra's head as she warily surveyed her daughter's prospective suitor.

With a curt nod, she motioned to Leon to join them. He carried himself with an almost military bearing as he strode over to their table. When he reached her and her mother, Edith was struck by his eyes, their intense shade of green.

She thought that he was one of the most handsome men she had ever seen.

Leon ordered a beer for himself and another *café turque* for Alexandra and one for Edith; the strong, sweet coffee was the only beverage the two women consumed.

There was small talk. Leon was adept at it, and Edith, given the chance, could be quite animated and charming—but Alexandra wanted it kept to a minimum. Relations between men and women were a business transaction, and beautiful young daughters were a valuable commodity.

Alexandra wasn't a practical person and never had been. But on that afternoon, she made the most pragmatic decision of her life: she resolved to set an exceedingly high price for Edith. She wasn't going to allow her child to become a passing fancy for this rich and elegant man. If he was serious—and that meant marriage—she would consider bending her seemingly unbending rule about dating, and allow her daughter to see him socially.

Leon had devoted himself in the last several years to marrying off each of his five sisters, so he knew the going rates for women. He had personally financed dowries for them, enabling them to find suitable mates—the pretty ones as well as the plain ones. All were given away in style, with lavish ceremonies and receptions paid for by him.

But there was a market for eligible men as well, and they, too, could command a high price. To date, no one had been able to meet Leon's price. He had remained resolutely single all these years, no matter how many eager matchmakers approached him, or how many prominent fathers tried to lure him with their attractive daughters and handsome fortunes, or how often his own mother pleaded with him to find a girl and settle down.

He had never even been tempted—until now.

The terms of the courtship were drawn up that afternoon, amid the intensely chic crowd—British officers in uniform, elegant ladies out for a day of shopping and enjoying a cool, refreshing *boisson* before heading back to their villas in Zamalek or Garden City or Maadi. There were different languages spoken at every table—French and English of course, but also Greek, Italian, Dutch, Armenian, interspersed with the occasional Arabic. It wasn't unusual for people to use two or three languages in the same conversation—even in the same sentence—because this was Cairo, and it was the most cosmopolitan city in the world.

Everyone was in a sparkling mood, especially Edith, who wasn't used to this kind of attention. She sat back and let her mother do most of the talking. But when Leon would question her directly, she'd light up and answer yes, she enjoyed teaching very much—especially little boys, they were such fun, *trés espiègle,* so mischievous. She adored the library. And what did he do? A businessman? Import-export? *La bourse?* Leon's answers were vague and intangible and impressive. Meeting this man excited Edith as much as being handed the key to the pasha's library—as if a new and magical world was finally within her grasp, a world of money and stature and position, as dazzling and luxuriant as white sharkskin.

Leon, for his part, was smitten. While he had known more than his share of worldly women, Edith was alluring precisely because she wasn't worldly. She was intensely captivating, though, and able to converse fluently in French and Italian and Arabic. The man who had steered clear of commitments his entire life decided then and there that he would marry this Levantine beauty.

The engagement was announced within weeks of the encounter at La Parisiana. Leon promised Alexandra that he would forgo the traditional dowry, since it was obvious the family had no means. He vowed to bear the costs of the wedding. Finally, he hinted that he was prepared to support Alexandra financially if she consented to the match.

Leon gave his intended a magnificent engagement ring known as a "cocktail ring" because of its elaborate mixture of rubies, diamonds, emeralds, and sapphires in a white gold setting. It was a bad omen that the ring disappeared days before the wedding. Félix, Edith's ne'er-do-well little brother, had stolen the ring and sold the stones. The wedding went on as planned, though the incident gave Leon pause.

The ceremony was held at the Gates of Heaven, the stateliest temple in Cairo. Afterward, the couple traveled by horse-drawn carriage to Jean Weinberg's photography studio to have their portrait taken. Weinberg, who had worked at the court of Atatürk, was the most talented photographer in all of Egypt. Yet even he was worried when he positioned the couple side by side for their official portrait. The bride was so small she risked being completely overshadowed by her husband.

Weinberg slid a small velvet footstool beneath Edith's feet. It promptly vanished under yards and yards of hand-sewn lace and satin.

The portrait came out so perfectly that Weinberg placed it at the center of his shop window in downtown Cairo, where it remained for months.

He signed his name in black ink, like an artist who has produced a masterpiece.

It is indeed a wonderful shot. Leon has abandoned his trademark white suits in favor of a classic double-breasted black tuxedo. He carries a top hat and white gloves and wears a sprig of lily of the valley tucked inside his lapel. Edith, a mist of dark hair framing her small, porcelain features, smiles slightly as she holds a large bouquet of white flowers—dozens of lilies and roses that trail from her hands.

The two are standing almost shoulder to shoulder.

It is an illusion, a scene from the same movie that opened at the café. No Selznick or Wilder could have scripted it better, this wartime romance. The last shot is of the couple embracing as they ride in their horse-drawn carriage around one of the most alluring cities in the world, a city touched by World War II yet at the same time shielded from its ravages.

Edith and Leon's wedding portrait,
signed by Jean Weinberg.

Except it wasn't a movie, because Leon and Edith were my parents. And the aftereffects of their café courtship would reverberate years later and thousands of miles away.

The account of their first encounter in the spring of 1943 would hold me in thrall long after I learned that much of what I'd treasured was make-believe, as illusory as their standing shoulder to shoulder in Jean Weinberg's photograph.

I held on to the glamorous image even when all glamour had gone from their lives and mine, and colonial Cairo was no more, and Jewish Cairo was a distant memory, and we had been banished to a string of shabby hotels in Paris and New York, until finally ending up in a corner of Brooklyn no wider than ten blocks, where thousands of other refugees from the Levant had also fetched up.

As we moved from country to country, from city to city, I learned to find solace in the fable of my parents' love affair. I would ask my mother to tell me of the romantic encounter again and again, I'd grill my father for details about that magical first meeting with Edith at La Parisiana.

"I find you very beautiful. Would it be possible for us to meet?"

What had drawn him to Edith? Why had he decided to marry her after spurning so many other women? Like so much of the lore that surrounded my father, Leon, each facet of his life took on a sheen and luminescence akin to the garments he favored, and I was never able to discern what was real and what had been artfully woven.

"Loulou, il faut reconstruire le foyer," my mother would tell me when I was a little girl; We must rebuild the hearth. It was a line from one of her favorite books, a work of fiction. At first, I didn't know what she meant. We were living in a cramped city apartment, not some country house with a fireplace.

Eventually, I came to understand that I was the chosen one, entrusted with the impossible task of taking our shattered family and our lost home and restoring them.

My point of reference became the photograph, so that I found myself always trying to recapture the promise of that wedding portrait, and more potent still, the vision of that dashing man in white sharkskin wooing the pretty dark-haired girl in a café in old, vanished Cairo.

BOOK I

THE CAPTAIN

• • • • •

CAIRO

1942–1963

CHAPTER 1

The Days and Nights of the Captain

On the first Thursday night of every month, Cairo grew completely still as every man, from the pashas in their palaces to the fellahin in their hovels, huddled by the radio and motioned to their wives and children not to disturb them. It was the night when Om Kalsoum, the Nightingale of the Nile, the greatest singer Egypt had ever known, broadcast live from a theater in the Ezbekeya section, her voice so transcendent and evocative that her fans could picture exactly how she looked as she came out onto the stage, enveloped in the lush white lace dress that softened and transformed her features.

This daughter of a village sheik had a cult following—porters and potentates, the intellectual elite and the illiterate masses, the beggars and the king—especially the king. But the most passionate audience for her songs about lost love and unrequited love and love forsaken weren't starry-eyed housewives but their husbands and brothers and grown sons.

To them, she was simply *al-Sitt,* the Lady.

She'd begin promptly at nine, fluttering her white voile handkerchief this way and that. Since each of her songs could last half an hour or more, her concerts went on well past midnight. "In the Name of Love," "What Is Left for Me?" "Tomorrow, I Leave," or her poignant classic "Ana Fintezarak"—"I Am Waiting for You"—they had heard these songs a thousand times, yet they still found them enrapturing, especially the verses that she would repeat over and over, each time with a slightly different inflection, a varied tempo, a changed mood.

It was the only night my father didn't leave the house or even his chair. He'd sit as close as possible to the radio, unable to pull himself away.

In the years before he met Edith, my father led the life of a consummate bachelor. He was rarely home, and when he left the apartment on Malaka Nazli Street he shared with his mother, Zarifa, and his young nephew, Salomone, it was not to return till dawn. His womanizing was the stuff of legend, as much a part of his mystique as his white suits, and there were countless other women before my mother, including, some whispered, the Diva.

Except for Friday nights, he didn't even bother to stay for supper. If he came back at all after work, it was to go immediately to his room and dress for the evening ahead, an elaborate ritual that he seemed to enjoy almost as much as what the night held in store.

He was meticulous and more than a little vain. He had assembled a wardrobe made by Cairo's finest tailors in every possible fabric—linen, Egyptian cotton, English tweed, vicuna, along with shirts made of silk imported from India. There were also the sharkskin suits and jackets he favored above all others, especially to wear at night. These were carefully hung in a corner of the closet, and if the local *macwengi,* or presser, dared to bring back a pair of trousers without the crease or fold exactly so, Leon would berate him and make him redo the job.

He always wore a diamond ring, and for the evening, he would add a tie clip in the shape of a horseshoe. White gold, encrusted with several diamonds, the clip was his good-luck talisman, and like all men who enjoy the shuffle of a deck of cards and the spin of the roulette wheel, my father was a firm believer in lucky charms.

His final act was to dab the eau de cologne Arlette on his hands and neck and temple. It was a popular, locally made aftershave with a fresh

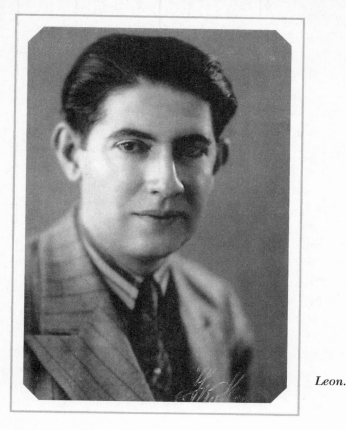

Leon.

citrusy scent that conjured the Mediterranean. Long after he'd left, the house still bore what the Egyptians would call, in their characteristic mixture of French and Arabic, *le zeft du citron*—the waft of lemon.

As he went out, Salomone, my teenage cousin from Milan, would poke his head from behind the novel he was reading to bid him good night, a tad enviously perhaps, and Zarifa would kiss both his cheeks lovingly but with some reproach in her magnificent blue eyes.

My grandmother came from Aleppo, the ancient city in Syria whose culture was far more rigid and conservative than Cairo's. She was troubled by her son's nightly forays and the fact he was still unattached and showed no desire whatsoever to settle down. Even now, in his forties, his restlessness continued to get the better of him. Until Edith, he never brought a woman home to Malaka Nazli, as that would mean she was the chosen one, and he had no desire to choose.

My father was a study in motion, taking long, brisk military strides early each morning to get from the house to his synagogue, then on to his business meetings, his cafés, and in the evening, his poker game and his dancing and his women. Because he tried to stay out of the house as much as possible, how convenient that his bedroom was at the front, facing Malaka Nazli, the wide, graceful boulevard named in honor of Queen Nazli, Farouk's mother. Because his room was only a couple of feet away from the door, he could slip in and out as he pleased.

Years later, I would hear that the lustrous lady of song, the devoutly Muslim Om Kalsoum, who was raised in a remote village where her dad had been the imam, had been my father's mistress. It was one of the many stories that persisted about my dad's prowess with women before and likely after he was married.

What I heard not simply about his womanizing but about every sphere of his life had a mythic quality, so outsize as to seem apocryphal. There was the fanatical devotion to religion and the hedonistic streak that compelled him to venture out in search of all that Cairo had to offer. There was the passion for clothes and food and women that made him a fixture at the leading restaurants and patisseries by day, and the cabarets, dance halls, and *cinémas en plein air* by night. Even his height and larger-than-life physique were cause for comment, as he was muscular and fair in a land of small, swarthy men.

The affair with Om Kalsoum had caused enough of a stir for word of it to spread in the family. It wasn't simply that a singer worshipped by millions had become involved with my father, because his ease with women was legendary. It didn't even seem incongruous that a star whose songs were all about the obsessive, indefinable aspects of love and desire would enter into a liaison with my obsessive, indefinable father.

Rather, what took everyone aback was that a devout Jew, scion of hundreds of years and consecutive generations of noted rabbis and scholars, would become involved with an Arab woman who was also a very pious Muslim. And perhaps as surprising, that a connoisseur of female beauty who didn't even deign to look at a woman unless she met his exacting standards would have a liaison with someone who, despite

Om Kalsoum.

her opulent wardrobe and finery, was rather a plain Jane. Her talent was the ultimate aphrodisiac.

My father's nocturnal wanderings would typically begin a block or two away from home, at the Bet el Om—the House of the Mother. That is how Farida Sabagh's home on nearby School Street was known. Tall and heavyset, Farida could no longer easily walk through her own doorway, which was always kept open. Yet her heart was as ample as her girth and she had an expansive, outgoing personality that made every-one want to come to the Bet el Om.

When Leon would arrive at her building, the porter, a simple peas-ant from the south, leapt to his feet. "Captain," he'd cry, raising his hand to his head in an awkward attempt at a military salute. Leon would smile, slip him a piaster, and continue on upstairs.

Farida didn't seem to mind the men who descended on her house night after night for rounds of poker. She liked to entertain her husband's friends and would stand there smiling and greeting them, "Etfaddalouh, etfaddalouh"—Welcome, welcome—before retreating to the kitchen to prepare some tasty treat that would sustain them for the intense gambling ahead.

Farida ran a rollicking household, with one hard-and-fast rule: bad news was absolutely forbidden. She wanted nothing to dampen her natural optimism, or spoil her typically euphoric, expansive mood, and the joy that reigned within. And so, if there were dismal developments about the war, and there seemed to be only dismal developments in 1942, she didn't want to hear about them.

Cairo was both protected from the Nazis' relentless march across Europe and Africa and profoundly affected by it. Tens of thousands of British troops were now stationed in and around the city. It was impossible to get a seat in one of the dozens of movie theaters because they were overflowing with British officers, and once the theaters emptied at midnight or one in the morning, the soldiers would move on to their favorite cafés, and stay nearly till dawn. After they'd had far too much to drink at La Parisiana and other watering holes, they lay sprawled on the sidewalks, and the military police were a familiar sight driving up and down the streets of Cairo, hoisting anyone who had passed out onto their paddy wagons.

To those who resented the colonial influence and wanted Egypt for the Egyptians, the English were a hated reminder of the foreign domination that had to end, war or no war. But for the Jews and foreigners who lived in terror of the Nazis, *les Anglais* were their only hope—their protectors and benefactors.

The war had even reached inside Malaka Nazli. My cousin Salomone had arrived in December 1937 from Italy, where Mussolini was about to make life impossible for Jews. His mother, Bahia—my father's older sister—had moved from Cairo to Milan in the early 1920s after marrying her husband. Italy had seemed filled with opportunities, and the family had thrived, but now they were stuck in a country that kept issuing ever-harsher decrees against its Jews, including racial laws in 1938 to rival those of Nazi Germany. Salomone had every intention of returning to Italy, at least for a visit, but the war made that impossible. He longed to be with his parents during their travails, but my father, exerting his indomitable will, forced him to stay put. Consoling himself with letters that flowed constantly between Cairo and Milan, Salomone focused on his studies, made friends, and tried to enjoy the comforts that life in Egypt offered, including the abundant love my grandmother Zarifa was more than prepared to lavish on her handsome, sensitive grandson.

My father was always amazed at the difference between his mother's house and the House of the Mother. In contrast to Malaka Nazli, which was quiet, contemplative, and somewhat mournful, the apartment on School Street was a constant party, with servants bringing plates filled with tasty hot kebabs and pitchers of cool lemonade made with fresh lemons.

On a typical poker night, most of the players folded by midnight and went home to their families. Leon wasn't ready to return to Malaka Nazli. He had the porter find him a taxi and headed to one of his favorite nightclubs, Covent Garden, the open-air dancing paradise, or Madame Badia's, the cabaret where Oriental men in tarboosh sat next to British officers in uniform and ogled the gorgeous belly dancers.

Madame Badia's girls were renowned for their beauty and skill. They performed in lush veiled costumes that evoked some Hollywood fantasy of the Arabian Nights. The Opera Casino opened in the 1920s, and Badia Masabni, its owner, a former belly dancer turned businesswoman, liked to showcase the finest dancers in all of Egypt. She had a good eye, and many of her young women were so talented they went on to become stars in Cairo's thriving movie industry.

It was possible to stop by on any given night and catch the delectable Taheya Karioka, a voluptuous brunette, perform the movements that made her the single most respected belly dancer in the Middle East—and a major draw for Madame Badia's establishment. Taheya managed to combine skill and sensuality on the dance floor as well as on-screen, as she starred in many movies. But her love life was also legendary and she seemed to collect husbands and lovers. She and the other girls danced with elaborate props that included a sword, a dagger, even a lit candelabrum they'd carry on their head like a crown as they swayed and gyrated and the audience held its breath. Ardent fans could also sit within touching distance of Taheya's archrival, Samia Gamal, who landed a role opposite Robert Taylor in the Hollywood extravaganza *Valley of the Kings,* and snared a Texas millionaire as one of her more than a dozen husbands. Some of the lesser-known girls were also fetching, and when they weren't dancing, they could mingle with the delighted clients.

For restless souls like my dad, Cairo was a palace of pleasure, a sybaritic Red Cross station. No matter how late he arrived at Covent Garden, which featured dinner and dancing amid acres and acres of manicured gardens, the orchestra would be playing continental melodies as hundreds of couples waltzed or did the tango or, for those under the spell of the British or the Americans, the swing or jitterbug. If he felt hungry, he could sit down as late as one or two in the morning and order a full-course dinner because this was Cairo, and the usual rules about when to dance and when to play and when to dine simply didn't apply.

The war hit home, of course. There was severe rationing in Cairo, and many necessities, like the gasoline that housewives needed to light the Primus and cook for their families, and chefs depended on to feed their hungry clientele, became exceedingly hard to come by. But there was also a flourishing black market, and those who knew the city intimately like my father were able to get their hands on crucial items such as sugar, oil, and soap. At Groppi's, the famed patisserie, it was possible to enjoy delicious pastries made with clarified butter as an orchestra played.

Despite all the shortages, the nightlife flourished and café society preened and the hedonistic streak of the city went almost unchecked. On a typical night, except for the abundance of soldiers out on the streets—British, of course, but also Australians, New Zealanders, South Africans, Indians, even some American GIs—as well as the refugees who had gathered from every corner of Europe, life could seem as glittering and self-indulgent as it had been in the 1920s and '30s, when Farouk's parents, King Fouad and Queen Nazli, reigned.

L'Auberge des Pyramides, which had opened in 1942, instantly overtook all of its rivals when it became known as Farouk's favorite club. It featured a ballroom under the stars, and on a good night, the king was almost certain to drop by with both an entourage and a determination to seduce the prettiest woman there, or whoever appealed to him the most.

The prospect of the king's arrival lent it a special cachet, though truth be told, L'Auberge was more gaudy than elegant—completely over-the-top in its decor, which came to include waterfalls, a swimming pool, and an indoor ballroom, in addition to its famed outdoor dance floor. Like every other nightspot, L'Auberge kept a table empty in case the pleasure-seeking monarch showed up unexpectedly. On a good night, Farouk would arrive and head straight to a table by the dance floor because it offered the best view of all the women in the room.

Woe to the man whose wife, girlfriend, or escort the king admired. The monarch was notorious for grabbing any woman he fancied, no matter if she belonged to another.

My father was such a habitué of the different establishments, there wasn't a club owner who didn't know him on a first-name basis. If there was a group of British officers—and there invariably was—he would join them at their table, and it didn't matter that he was both an Arab and a Jew. He was really one of them.

"Captain," a voice would cry out in the darkness, "Captain Phillips," and Leon would scan the familiar khaki uniforms and smile, knowing he was among friends. He would join the officers for a drink, and banter in English, affecting an accent that was almost as posh as theirs.

They liked my father because he bridged both worlds—he could play the part of the Captain to the hilt, displaying the poise and polish of someone to the manor born, suggesting an education and breeding he'd never really had.

Sometime in the night, he would treat them to a favorite party trick: he would offer to read their palm, because that was what Egyptians were supposed to know how to do. Taking their hand in his, he would remark how their life line was especially long and that defeat of the enemy was in sight.

He was relentless about meeting women. It was his great diversion, equal to—and at times surpassing—gambling. But he'd rapidly lose interest and move on, in search of new quarry, new opportunities. His reputation as a *flirteur* only added to his fascination. If he met a woman who interested him in the course of these evenings, he'd pursue her ardently. But he'd also disabuse her of the notion he was available for more than a passing dalliance. The thrill for him was entirely in the hunt, the courtship, the chase.

When the evening was over, he'd wander alone back to Malaka Nazli. Only his bride or his intended could ever be permitted to enter his mother's house. He was never seriously linked with anyone. And, truth be told, some of the alluring, worldly European beauties he courted would have been astonished to see that he resided with an elderly, autocratic Syrian woman who fussed over him as if he were a cross between a god and a child.

As he pushed open the door, he'd see my grandmother at her usual guard post—a small hard chair in the living room, a kerchief on her head. She hadn't been able to fall asleep, she'd explain to him in Arabic, because she was so worried about him, fretful about his wanderings across the city. He'd kiss her tenderly on both cheeks, then retreat to his room.

No matter how much he devoted himself to her care, Leon was entirely immune to her words of reproach. Zarifa could only express her anguish and disapproval by sitting up night after night on that hard-

Zarifa of Aleppo.

backed chair, her long white hair loose on her shoulders, her blue eyes welling up with tears, waiting for the key to turn and the door to open and her son to reappear. Only then would she allow herself to go to sleep.

Still heady from the night, my dad would linger awake in the privacy of his room. He'd carefully remove his jacket and silk shirt and place his jewels and tie clip and cuff links in a special box. He took exceptional care to fold his shirt and hang up his suit so they could both be worn again. It was almost dawn, which meant a new day—and a new night—were on the horizon.

MY FATHER PROSPERED AS a businessman in Cairo, but no one could figure out how he made his fortune or even say what he did, exactly.

He was certainly an avid investor, with a passion for *la bourse,* the rough-and-tumble Egyptian stock market, but there were also his skill as a broker, buying and selling products that ranged from plain brown wrapping paper and cellophane to food additives, sardine cans, and complex pharmaceuticals; the grocery business he once co-owned with his older brother, Oncle Raphael, specializing in the purest olive oil and the finest cane sugar; his expert knowledge of textiles, especially Egyptian cotton, one of the most desirable fabrics in the world; and his frequent trips to Alexandria for his import-export business, though what he imported and exported is unknown. At some point, he began to do business with the exciting American soda conglomerate that was setting up a beach-head in Cairo—Coca-Cola. He provided some of the key ingredients used to produce the famous soda.

Even those who watched him firsthand, like his nephew Salomone, weren't too sure of the nature of his ever-changing business.

All that is clear is that Leon never held a real job. He never collected a steady paycheck save once, as a teenager, when he briefly went to work for a bank. En route to becoming *un banquier,* one of the most prestigious occupations in all of Egypt, he found that he couldn't endure sitting at a desk and hated the hours, the routine, and above all, the need to report to other human beings and be subject to their wishes and whims.

The Captain could never allow anyone to give him orders.

Instead, he struck out on his own. As a young man coming of age in British-ruled Cairo, he made himself indispensable to the colonialist powers. He had to overcome two hurdles—the British dislike of local Egyptians, whom they called "wogs," and their distaste for Jews. Gifted with languages, he mastered seven—English, Arabic, French, and Hebrew, of course, as well as Italian, Greek, and Spanish. This enabled him to function as an interpreter, guide, and go-between, and he could take his British friends to the most obscure parts of Egypt and help them communicate with the most intransigent local characters. In a way, this was his first stint as a businessman, when he became a broker and middleman between two worlds—cosmopolitan colonial Cairo and mystical, sensuous Islamic Cairo.

There were periods of great prosperity, while at other times he struggled. But he had learned from my Syrian grandmother to keep both his good fortune and his misfortunes to himself, and never, ever showcase his wealth: that was the legacy of Aleppo, the ancient Syrian city where Zarifa and her husband were born and had fled shortly after the turn of the century, along with their ten children, including my infant father. It was a period of turbulence, when many of the Jewish families who had lived in Syria for centuries packed up to go, fearing economic privation as well as religious persecution.

Tragedy had struck after the family had settled in Cairo. Leon's father died after a hernia operation. Shortly thereafter, there had been the death of his sister Ensol, the beauty of the family, and her husband, who were either murdered in a train speeding from Cairo to Palestine, their throats slashed by an unknown assailant, or killed in an accident. The tragedy was never talked about, for that was the Aleppo way, but every once in a while my grandmother would cry out, "Ensol, Ensol," to no one in particular. She and my father had taken in their children and helped raise them, but the bad luck continued. Ensol's son had gone insane and remained confined year after year in the Yellow Palace, the vast jasmine-scented lunatic asylum located in Abbassiyeh. Another blow came when Salomon, Zarifa's second oldest son (not to be confused with his namesake, Salomone, who came to live with them from Milan), returned home from the Collège des Frères, the prestigious Catholic school to which even devout Jewish families sent their children, and announced he was converting and entering the priesthood. To a family whose ancestors had been the religious leaders of Aleppo for hundreds of years, the defection was both heartbreaking and incomprehensible, a mystery they would ponder all their lives.

My grandmother mourned him as if he were dead: in old Cairo, that is what you did with someone who left the faith. Zarifa recovered, yet she never ceased to talk about the old days and the old ways, reminding her children and grandchildren of the family legacy. Even in cosmopolitan Cairo, she insisted on following the ways of "Halab," as she referred to Aleppo in Arabic.

Barely educated—girls rarely were, in Syria—my grandmother only spoke Arabic, and when she went out, she'd don a *chabara,* a long, lustrous black robe favored by Arab women that covered her hair and body and reached down all the way to her ankles. She loved to say how in Syria, the family had dined with kings. She never elaborated, and it wasn't clear if there had actually been monarchs in her social circle in Syria or if she was simply referring to the family's illustrious past, the time when the family name was revered for the generations of rabbis it had produced and the religious texts they had authored.

Though Aleppo was long ago, its culture still exerted a powerful, almost mystical hold on all those who traced their origins there, and always would, whether they lived in nearby Cairo, or settled in far more distant capitals—New York, Buenos Aires, São Paulo, Johannesburg. To be a Halabi Jew meant obeying a set of social and religious conventions that dated back centuries and almost never changed with the times—rules that spelled out precisely how to live and how to die, how to worship, marry, raise children, and of course make money, because Aleppo's culture was profoundly materialistic, and wealth mattered second only to God and family.

Leon was so reserved that no one was ever entirely sure when he was faring well or when he was close to bankruptcy, and that too was a residue of Aleppo, where a man was supposed to trust only his immediate family—blood ties mattered above all—and even they were to be kept in the dark about his work.

Mostly he did well, except for the Depression, when the grocery business he ran with Oncle Raphael went under.

Aleppo was also a secretive, almost paranoid culture so that despite being gregarious and sociable, Leon was fundamentally a solitary man. He was always moving. Though he could have taken horse-drawn carriages and taxis to his business meetings, he preferred to get around on foot, maneuvering briskly through streets that often weren't paved, that were barely more than dirt roads.

He could traverse distances that would exhaust far younger men; even in his forties, he had remarkable energy and exuberance, and though his favorite outlet was the dance floor at night, he enjoyed wandering for miles around Cairo in the early morning, approaching clients when the streets were still and there was a slight breeze, before the air became heavy and the city shut down because it wasn't humanly possible to work in the heat of an Egypt afternoon.

Back in the late 1930s, when he was still trying to rebuild what he lost in the Depression, he took his nephew to work one summer day. Salomone was a strapping young man, almost as tall as Leon, and about twenty years his junior, yet even he couldn't keep up as they walked and walked in the scorching Cairo sun, paying calls on more and more clients. Leon would approach small vendors—simple fellahin selling juice from stalls the size of a large box. He'd converse with them, and he wouldn't be at all patrician. On the contrary, he would transform himself into a man of the people like them.

Once he had cemented a bond, he would whip out his notebook and carefully take down their orders:

Ten bottles of soda. Eight cans of sardines. A dozen bars of soap. Four large kegs of olive oil. Two sacks of flour.

To my Milanese cousin, it all seemed so slight and insignificant. Was it possible that his glamorous, aloof Oncle Leon, who was always leaving home for some alluring world beyond Malaka Nazli, was so small-time? That he made his living collecting a few piasters from the sale of a couple of jugs of olive oil and a dozen sardine tins?

Except that several hours and countless merchants later, Leon was still going strong, jotting down more orders, whereas Salomone was on the verge of collapse. He had entirely lost track of what Leon was doing, and had no idea of the sheer volume of his business, and wanted only to get out of the sun and go lie down in his dark, cool room in the back of Malaka Nazli.

Within a few years, Dad had graduated from the little grocers near Malaka Nazli to customers that included the largest concerns in Cairo, including Groppi's, the legendary Swiss patisserie that was a central meeting point for café society, and Spatis, the Greek soda manufacturer whose bottles of fizzy, lemony pop were all the rage in Egypt, and later still, Coca-Cola. He thrived on his reputation as a *négociant*, a broker, because he could locate any product, however trivial or obscure, and he was known as a man of his word.

He profited handsomely because he was so unencumbered, with almost none of the trappings of a big businessman—no capital, no overhead, no inventory or backlog, no warehouse, no discernible assets (or liabilities), no employees, and most important of all, no boss.

He didn't even maintain a line of credit with his suppliers; he operated on a system of hard cash and absolute honor. Whatever he bought, he paid for then and there. He disdained contracts and signed documents, and when he gave his word, it was enough; he had a special fondness for what he called "gentlemen's agreements."

One of his clients was a purveyor of gourmet products, including spices, canned goods, and culinary supplies. My father would visit the firm on El Azhar Square, by the large mosque, at least once a week. As he swept through the office on his way to meet with the top managers and place his orders, he was careful not to overlook anyone, even lowly clerks. He'd reach into his pocket and fish out some bonbons and fling them on every desk along the way, as if he were throwing dice on a gambling table.

He filled a room; the entire staff would stop what they were doing and look up. And unlike most clients who dropped by and chatted familiarly in Arabic, Dad was fond of dropping Britishisms—offering a clipped "Good morning" or "How are you, old chap?" He'd pepper his conversation with "Jolly good," and some of the staff would call out, "Captain," and rise and offer a playful salute.

The company was essential to his business, yet here, too, while he was a sizable client who placed frequent orders, my father never kept an account as so many others did, which would have allowed him to enjoy a line of credit and order as much as he wanted. He preferred to buy only the items he needed—*sel du citron*, used in baking, for instance—and pay cash, removing wads of bills from his brown leather wallet.

He had no desire to incur debts, and besides, cash transactions left no paper trail. Nearly all what he did was off the books, and he conducted his business affairs the way he conducted his personal affairs—in strict secrecy, confiding in no one. It was what made him a superlative businessman, and it was the Aleppo way. How he turned a profit ranked among the mysteries of the universe—as elusive as the soft gleam of his jacket, as indefinable as his charm.

ON FRIDAY NIGHT, HE stopped. He was a different person on the eve of the Jewish Sabbath, arriving well before sundown with a large bouquet of roses for my grandmother and embracing his nephew before preparing for synagogue.

Zarifa would be in the kitchen, more cheerful than usual because this was the one night of the week she would have her son at the dinner table. She prepared the meats and chicken and rice that were her specialty, cooked in the style of Aleppo, with a hint of fruit. There was also stuffed eggplant—white, not black, because black brought bad luck. Every self-respecting Levantine household was obsessed with warding off the evil eye. On the Sabbath, you avoided dark clothing, dark thoughts, and dark foods. Even black olives were banned from the table to guarantee a good week.

Dressing for temple required the same attention to detail as dressing to go out on the town. Leon put on a white shirt made of the finest cotton and a white jacket. For jewelry, he liked a more sober tie clip adorned with a single pearl.

By sunset, the streets around Malaka Nazli welled up with men in elegant suits holding velvet or satin pouches as they walked. They were on their way to temple, and the satchels contained their prayer shawls and prayer books and skullcaps. In a way of life the world has now forgotten, this quintessentially Arab city was supremely accepting of its Jewish inhabitants. Muslims and Jews lived in close quarters—in the same streets, the same buildings—and usually very harmoniously. No one wore the skullcap outdoors, it is true, religion was a discreet affair—yet it was obvious to anyone that these were Jews on their way to pray at one of the dozen synagogues that flourished around our Ghamra neighborhood.

My dad had his pick, and depending on his mood, he would pray in this temple or that. He relished a small, simple, deeply intimate house of worship in an alleyway a couple of streets away known simply as "le Kottab," or the Schoolhouse. Another favorite was *Ahavah ve ahabah,* the Congregation of Love and Friendship, where his favorite rabbi, a diminutive hunchback named Halfon Savdie, drew a lively crowd because of his eloquence and amiability. Dad adored Rabbi Halfon.

When services were over, the exquisitely dressed men once again crowded the streets, laughing and joking as they hurried home to their wives and children, anxious to sample the special Friday-night cooking whose smells filled the night air of Ghamra.

At home, Zarifa had set the dining room table with a white tablecloth, and after my father and Salomone were seated, she and the maid came in with the courses she had devoted much of her day to preparing, since it was necessary to cook both for Friday night and Saturday. Sabbath dinner was a quiet affair, with no cousins or guests or relatives.

My father poured a cup of homemade wine from a bottle. Everyone stood up as he recited the blessing on the ersatz wine, and afterward he and my cousin went to my grandmother and kissed her hand, then returned to their seats.

At last my grandmother smiled, delighted to have her tall, handsome son on one side of her, and her dashing grandson on the other. They no longer dined with kings, but she was in the company of two princes.

Come sunset on Saturday, the well-dressed men with velvet pouches under their arms could be seen returning from temple. The Sabbath was over, and each held a green sprig in their hands. It was a branch of myrtle, which emits a powerful, almost dizzying scent. They had been given the pungent herb at the close of the service, when the rabbi recited the Prayer of the Perfume. A new week was beginning, and the soul needed to be fortified.

What none of Leon's acquaintances could do was reconcile the man at synagogue, who seemed so immersed in his prayers, and was relent-less about observing every ritual, fasting every fast, and obeying every possible commandment, and the man who disappeared night after night for forbidden, sinful pursuits.

It was as if two people resided within one sharkskin suit, one who was pious and whose vestments resembled those of the priests at the Great Temple, all white and sparkling and pure, and the very different creature who led a secret, intensely thrilling life far beyond Malaka Nazli.

It was a life of pleasure, and one that Leon's mother disapproved of. She was insistent it had to end. Under Aleppo's code of honor, a man was granted enormous latitude as to when to marry; unlike a woman, he could be in his twenties, thirties, forties, and still make a dazzling match. But under that same code, he, too, had to choose a wife. There were no permanent dispensations, even for men like my father who had no desire whatsoever to settle down. He had ignored Zarifa in his youth, but her entreaties, though he'd never admit it, were beginning to wear him down.

ONE TERRIBLE MORNING IN the summer of 1942, Rommel boasted on the air he would be at Groppi's by five that afternoon. Mussolini, mean-while, sent word that Egyptian women should prepare their loveliest party dresses for the bash he planned to throw once the Nazis con-trolled Egypt. The Nazi general was less than an hour away from Alex-andria, his seemingly unbeatable army ensconced in the town of El Alamein, and he was confident he would be able to vanquish Cairo in time to enjoy afternoon tea at Shepheard's and a pastry at the legendary patisserie. As if that would establish the Reich's supremacy.

Groppi's had as much symbolic value for Cairenes as Maxim's and the Hotel Meurice for Parisians, and the Nazis had made a great show of taking over both after invading Paris. The Germans had a fine-honed instinct for grabbing a city's most desirable properties and then settling in with great fanfare, and in Cairo, the café and pastry shop on Sulei-man Pasha Square and its annex on Adli Pasha were unquestionably the places to see and be seen. Indeed, they were the favored meeting ground of all the Reich's enemies—the French, the British, the Australians, the Greeks, and the Jews.

For Leon, if there was a paradise on earth, its name was Groppi. Part café and part patisserie, part bar and part trysting ground, the fabled establishment was never as glittering as when British officers stationed in Cairo decided to make it their personal haunt in the war years. The officers loved to escort their wives—or their mistresses, or their girlfriends—there for a spot of afternoon tea and dessert. And because the British were there, so was my father, and because they adored it, so did he.

Every Jewish household was panic-stricken at the prospect of the Nazis storming Egypt. "The Germans will arrive and cut our throats," they said.

At the House of the Mother, Farida Sabagh had been telling the men for days to stay calm. Cairo was after all the land of prophets and mystics, she reminded them all—the birthplace of Moses, the home of Maimonides, the city where Jeremiah, the mournful prophet, and Elijah, the immortal one, were known to have sojourned. They would never abandon us, Madame Sabagh proclaimed, raising her arms toward the sky in prayer.

"Dieu est grand," my father cried out, eyeing his new hand of cards.

As a gambler, Leon believed that luck could change, and that the British were due for some good luck. But as a close observer of the war, he feared that Rommel, the German "Desert Fox," was unstoppable, and that only a miracle could save Egypt. As a devout Jew, he believed with perfect faith in the possibility of such a miracle.

The men continued playing, reassured by the notion that holy men of centuries past were watching over them. All the Jews of Cairo felt vulnerable, especially those who had already escaped from Hitler's clutches in Europe and were now frantically making plans to flee again.

In her kitchen on Malaka Nazli, my grandmother wept softly over her Primus. Zarifa always cried when she was afraid. She had a daughter in Nazi territory and now the entire family was at risk, for there was no doubt what would become of the Jews were the Germans to prevail at El Alamein.

It quickly became clear that Rommel wouldn't get to savor even one morsel of Groppi's fine pastries. He suffered the worst defeat at El Alamein that the German army had known to date, and was forced into a humiliating retreat, his army in tatters, his tanks destroyed, and his legendary pride shattered. Cairo breathed normally again, and Cairo's *haute societé*—British officers, Frenchmen, Australians, and of course, Jews—converged on Groppi's to celebrate, confident for the first time the war would go their way. At the afternoon tea dance my father joyfully attended, the orchestra played on and on.

While the Nazis had been routed from Egypt, and were suffering one humiliating defeat after another in North Africa, the rest of the world was still at war. Refugees continued to pour into Cairo, seeking haven. Everyone was making fast decisions, as if there was no time to waste. There was also the fact that in Egypt, at least, the Jews had stayed miraculously safe.

As Cairo's was one of the few Jewish communities to be left whole, it was now possible to envision a future, to think of marriage and children. In the spring of 1943, several months after England's stunning show at El Alamein, and as it routed the Nazis out of Africa, my father brought home a young woman with a regal bearing and the wide, frightened eyes of a doll. No one was more surprised than Zarifa when he announced his engagement to twenty-year-old Edith.

CHAPTER 2

The Season of Apricots

Alexandra and Edith made their way to Malaka Nazli for the first crucial family get-together. Mother and daughter tried their best to get their bearings in this house where the matriarch dressed and behaved like an old Arab woman, so that even in the heart of Cairo, it was as if they had wandered into another culture, another era. Lending them moral support was a man they worshipfully called Oncle Edouard, though he was in fact Alexandra's stepson, and Edith's half brother. Edouard was older, richer, and formidable looking—the closest Edith had to a father figure since her own father had abandoned them. He also cared deeply about her and came prepared to do what only a man could do, offer his protection and negotiate any financial details of the impending union.

In truth there wasn't much to negotiate: Edith had nothing, no financial assets whatever.

My mother didn't have a clue, of course, about Leon's lifestyle. She didn't know of his habit of leaving Malaka Nazli early each morning and not returning till after midnight, and the whirlwind courtship and family get-together didn't give her the opportunity to learn. The moment they walked in, Zarifa praised her son's fiancée for her beauty, calling her "heloua, heloua"—lovely, lovely—and patting her gently on the head. She disappeared into the kitchen and emerged with platter after platter of elegant handmade pastries and cookies, fruits and nuts, more pastries, and large carafes of freshly squeezed lemonade.

The wedding would take place almost immediately.

Of course, the ceremony would be at the Gates of Heaven, the large synagogue downtown. It was the only temple that could possibly accommodate all the aunts and uncles, cousins, nephews, and nieces on Leon's side.

Edith shyly showed off her engagement ring, and her future mother-in-law bent down to peer at it, almost as if she were at the famous old souk in Aleppo inspecting a lemon or an apricot, and pronounced it satisfactory.

Watching on the sidelines was Salomone. He didn't say a word beyond the initial introductions, but he was struck by Edith's youth and vulnerability—she couldn't have been much older than him, yet there she was, marrying his uncle. Salomone would have to give up the spacious master bedroom he had enjoyed in the back. It was the room Edith and Leon would be occupying after their honeymoon.

When the newlyweds returned from Ras-el-Bar, the seaside resort whose charming bamboo huts overlooked the Mediterranean, my father seemed to have turned over a new leaf and to have abandoned most of his old habits. He still left early in the morning to go to synagogue and then to work, but instead of staying out all hours as before, he made it a point to come home for lunch and again at dinnertime.

He'd flick on the large wooden radio. He seemed content to while away the evening reading the paper and listening to the broadcast of 'Om Kalsoum singing at a downtown club, a reminder of the life he had left behind.

My grandmother immediately lauded her new daughter-in-law as lovely and refined. Zarifa appreciated the fact that Edith was learned—teaching was a very prestigious occupation—yet in truth, education mattered little to the elderly Syrian woman who could barely sign her own name. What counted were Edith's attributes as a wife, whether she was a capable mate for Leon and would—*Inshallah*—bear him many children.

It was obvious within weeks of moving in that Edith lacked many of the necessary qualities, except youth and beauty.

To my grandmother's dismay, Edith showed little interest in the affairs of the house, and even less interest in cooking. She seemed nervous and uneasy in the one room that consumed and preoccupied Zarifa—the ancient little kitchen. My mother tiptoed curiously by the Primus, with its small sputtering sounds and heady scent of petrol, approaching it as if it were a foreign object. Alexandra, Edith's mother, was a different breed entirely than her new mother-in-law, and it wasn't only that she spoke French and Italian and disdained Arabic. She had an aristocratic outlook and demeanor, the product of a privileged upbringing in a wealthy Alexandrian Jewish family, and she had raised Edith to regard housework as beneath them even when there was no longer a hint of privilege in their lives. She encouraged her daughter to read books or go to the cinema, not waste her time in the kitchen.

After her father had left them, Edith had assumed most household duties and tried to care for all three of them—Alexandra, her brother, Félix, and herself. Yet she had inherited Alexandra's disdain for "women's work." Besides, Edith assumed that since she had married into a family of wealth and stature, servants would do all the work.

This was the mind-set, fanciful and disastrous, that my twenty-year-old mother brought to her marriage to my forty-two-year-old father. If not for the maid and Zarifa's intense personal involvement in the affairs of the house, the beds would be left undone, the cupboards would be bare, the floors would accumulate dust, the Primus burner would remain shut, and no guests would join the family because there would be little to serve them.

Edith wasn't lazy or incompetent; she was simply terrified of the autocratic old woman who was the true ruler of Malaka Nazli. She was only too willing to cede control of the kitchen to her mother-in-law. She preferred to stay in the quiet master bedroom devouring her latest novel, and she missed her job at the École Cattaoui, which she'd given up—regretfully—because no self-respecting bride in 1940s Cairo continued working when she had a husband who could support her.

Zarifa issued her verdict about my mother: she is not a good home-maker. It was like a death sentence in a community where a woman is judged first and foremost by her appearance, and second, by how she runs her house. News of my mother's shortcomings spread beyond Malaka Nazli, to the homes of Zarifa's daughters and daughters-in-law.

The family was curious and by nature gregarious. They would drop in—my father's sisters, Leila, Marie, Rebekah, and his older brothers, Raphael and Shalom. The only ones who didn't come were Joseph, the oldest, and the richest, who lived in a mansion in Zamalek and was estranged from the family, and Salomon, the priest, who lived in Jerusalem and wasn't welcome on Malaka Nazli in any event.

All were somewhat puzzled by their brother's choice of bride—Edith was very pretty, certainly, but so slight and unassuming. After all these years, they had expected someone more prepossessing, not this poor humble girl from Sakakini.

Maybe they could tell she was no match for the Captain. In a society where men ruled, a woman had to have a strong character to hold her own in a marriage. It was clear to everyone that for all her attributes—her beauty, her charm, her education, her sense of humor—my mother would never be able to stand up to Leon.

Edith, for her part, liked most of her in-laws, particularly Rebekah, who had a good heart and seemed anxious to make her feel welcome, and Oncle Shalom, the poorest member of the family, who limped terribly in his elevated shoes, but made up for his infirmity by being exceedingly kind. But she was surprised by the women's custom of wearing robes or dressing gowns when they sat down at the dinner table. She was too shy to ask them why—was this some sort of quaint Syrian tradition brought over from Old Aleppo? She didn't feel comfortable imitating them. She wore a dress at dinner, whether there were guests at the table or not.

Then, there was the food itself. Lavish meals were served at Malaka Nazli almost every day, and every single dish seemed to be cooked with fruit, mostly apricots. It was all so different from the foods Edith was used to eating, made with onions and leeks, drenched in lemon, and fragrant with garlic. The tastes and smells from Zarifa's kitchen added to Edith's sense of herself as a stranger who had wandered into a foreign country named Malaka Nazli.

MY GRANDMOTHER ZARIFA BELIEVED that certain foods possessed magical powers.

Bananas and raw eggs were at the top of her list, along with almonds, sour cherries, olives, and, above all, *mesh-mesh,* ripe, luscious apricots. Zarifa would slip them into virtually any dish that was simmering on her Primus, and she was such a gifted cook that every course she prepared, from simple grilled meats to elaborate stews and roasts, emitted a mysterious, vaguely fruity aroma.

Anyone who came to dine at Malaka Nazli could count on a transcendent experience. What was in there? they'd wonder, amazed at their sense of absolute well-being.

They'd press their diminutive hostess, but she only smiled and refused to answer. But over coffee, she would drop a hint or two about a charmed and distant past.

"Once upon a time, we dined with kings," my grandmother would exclaim in Arabic. Her soft voice rose as she stirred her coffee with a rapid encircling gesture, using the small gold spoon that no one else was allowed to touch.

In the summer there was news that brought a smile to the old matriarch: Edith was expecting. At last, Leon would have an heir.

But there was also tension at Malaka Nazli. The joy and excitement over the baby was overshadowed by a sudden turn in the marriage. Without warning, my father had gone back to his routine of going out night after night and never telling anyone—not even his new wife—where he was going. His old lifestyle resumed.

They had barely been married a few weeks, yet it was as if nothing had changed.

He moved out of the master bedroom he shared with Edith and reclaimed his old room, steps away from the front door. My mother was bewildered. She had a young woman's pride: Hadn't she been the *only* woman able to snare the elusive boulevardier? It didn't help that she was now pregnant, frightened and anxious about giving birth. Worst of

281-Malaka Nazli.

all was that she had no one to confide in, certainly not in her mother-in-law. Even Alexandra, who'd had her own disastrous experiences with a husband, had encouraged this union, seeing an opportunity for her daughter to marry a man of means and stature, and disregarding the obvious danger signs—the large age difference, the immense cultural chasm between an older, worldly gentleman and a meek, inexperienced young woman.

And while there was Salomone, she could hardly speak to him about her marital woes. Thrown together in close quarters, they were genuinely compatible. Both were intellectual, and Salomone shared Edith's passion for books. She was comfortable conversing in Italian, which was a relief to the young man, who missed the language and ways of his native country. Yet she couldn't tell him what was paining her.

In a way, she didn't have to: there were no secrets on Malaka Nazli.

Salomone had watched the entire drama unfold—the hopeful engagement, the refined young fiancée, so different from the other women in the Syrian clan, the promising start of the marriage, followed almost immediately by my father's return to his old ways, and my mother's despair about being abandoned night after night.

Leon never took my mother along when he left, and he never answered her questions as to where he was going.

With her Proustian imagination, her knowledge of the seamier side of life that came only through literature, Edith feared the worst. The bon vivant qualities that had attracted her took on a darker cast. Surely my father had a mistress, she thought, or more than one. What else does a man do, far from his family, at 1:00 a.m., in Cairo?

Her vast readings couldn't provide her with the insight a more worldly woman would have had, and because she was fundamentally timid, in no way a match for this man who was answerable to no one, she found herself trapped in an unhappy union only months after it had been consummated.

There was no point in complaining to Zarifa. No self-respecting Aleppo matriarch would ever take the side of a daughter-in-law against her own son, at least not publicly.

In fact, though she defended Leon, my grandmother was deeply saddened by the turn of events. She, more than anyone, had hoped that her son's restless nature would be quieted once he was married. Yet even a beautiful, desirable young wife and the prospect of a child seemed not to have altered my father's ways in the least, and he insisted on coming and going as he pleased from Malaka Nazli.

NINETEEN FORTY-FOUR BEGAN WITH some promise. The showdown at El Alamein two years earlier had proved a watershed, halting the German army's relentless advance into North Africa.

"The Battle of Egypt," as Churchill affectionately dubbed it, marked a turning point in the Allies' fortunes. In one fell swoop, the British had dashed Germany's dreams of seizing Cairo and controlling Egypt and the Suez Canal.

A jubilant Churchill, who visited Cairo several times during the war, declared, "Before Alamein, we never had a victory, and after Alamein, we never had a defeat." Egypt, the land of miraculous unfoldings, had brought the Allies luck.

But the war wasn't over, and the war against the Jews had only intensified. It was always jarring to speak to the refugees who passed through Cairo. They were so breathless, on their way to the southernmost reaches of Africa, because anyone who had managed to flee Europe couldn't get far enough away and felt they had to keep running to the ends of the earth.

There had been no news about Salomone's family in Milan for months. The heavily censored postcards from his parents that had arrived early by courtesy of the Red Cross, and offered some reassurance, had stopped coming altogether sometime in 1943. Despite the dearth of information, Salomone tried to keep his fears in check.

He went about his business as if all were well in the world. After finishing the lycée, he had landed an excellent job as an accountant. Each morning, a chauffeured car would arrive to take him to his office, and would drop him off in the evening in time to dine with Zarifa and Edith and, if he was around, Oncle Leon. On weekends, he kept busy with an active social life. Tall, slender, and impeccably groomed, Salomone was proving to be popular with women. He had my father's boulevardier tendencies, including a love for fine clothes, good food, and attractive women. Matchmakers began approaching him the way they once had his uncle.

Young Salomone,
Alexandria, 1939.

If only there were news about his parents and his sister.

A portrait of Salomone's mother surrounded by her four children when they were toddlers, smiling and exquisitely dressed, occupied a place of honor in the dining room for all visitors to see—a young woman so intrepid that when she became engaged to his father, a man twenty-five years her senior, she seemed undaunted by the prospect of leaving her family and comfortable life in Egypt to settle in Italy.

After eight years in her house, Salomone had adopted many of my grandmother's belief systems, especially concerning the importance of food. Salomone learned that apricots were the fruit of God, while almonds and other nuts had medicinal powers. For everyday ills, coffee was the all-purpose cure.

That is why, when he woke up feverish and under the weather a day or two after celebrating the New Year, my cousin decided to head by trolley directly to À l'Américaine, the jaunty Groppi annex whose frothy cappuccino was the best in all of Cairo. But on this January morning, his teeth were chattering and he was sweating profusely, and even the comforting Italian coffee that reminded him of home failed to have its intended healing effect.

He staggered out and hailed a taxi back to Malaka Nazli. He was greeted at the door by my father, who had chosen that afternoon to stay close to his bride, who was nearly eight months pregnant. When he saw his nephew, feverish and pale, wheezing, at times unable to breathe, Leon ordered him to go to bed immediately.

The family doctor, who made house calls any hour of the day or night, came immediately, examined Salomone, and expressed alarm at his fever. But he was unable to make a diagnosis. My father summoned another doctor, and then another; none could say for sure what was the matter.

It was time for Salomone to consult *un spécialiste.* In Cairo's pantheon of professionals, doctors were revered but specialists occupied an exalted place at the top. Unlike their colleagues, they never made house calls but expected their patients to make their way to their private offices in the tonier sections of the city. None was more renowned than Dr. Grossi, an Italian pulmonologist who had made Cairo his home, and was considered the finest lung specialist in Egypt.

My father bundled his nephew in a warm coat and ordered the porter to summon a taxi. With Leon's arm around him, Salomone made the painful journey over to Emad-Eldin, the fashionable district where Dr. Grossi maintained his private practice.

Calmly, methodically, Dr. Grossi examined my cousin. After administering multiple tests, including a crude chest X-ray, he was confident in his diagnosis: Salomone had *la pleurésie,* an inflammation of the heart that was sometimes fatal, he decreed. In this era before antibiotics were widely available, pleurisy was especially difficult to treat. Still, Dr. Grossi was certain that he could devise a course of treatment that would work.

First and foremost, he ordered complete bed rest. Salomone had to stay in his room, as motionless as possible.

The second aspect of the treatment involved food. My cousin was ordered to eat—to eat constantly, as much as he could handle. There were no medicines or potions Dr. Grossi could offer in a country that hadn't even seen penicillin yet. If my cousin had any hope of surviving this deadly infection, he had to consume a high-calorie diet rich in calcium and minerals.

Though they didn't meet, Dr. Grossi managed to endear himself to my grandmother as few men of science ever would. He confirmed what Zarifa believed with every fiber of her being: that food was the weapon of choice in combating even the most complex ailments.

"Once you get home, don't move," Dr. Grossi reminded my cousin as he painfully sat up on the examining table.

Zarifa was waiting for them in the living room, sipping a cup of Turkish coffee flavored with a hint of orange water. She stirred the cup anxiously, round and round, with her gold spoon. After Leon, Salomone was her favorite person in the world, dearer even than her other sons and daughters.

Then and there, my father decided to move out of his bedroom and grandly offered it to my cousin. It was the most pleasant room in the house, with two beds and a large window that faced Malaka Nazli. He could rest even when his bed was being changed: he could merely roll over to the other bed.

Salomone had never felt so sick as in those first weeks of 1944. He slept a great deal, and when he didn't, he read obsessively. As it happened, in a number of the novels the hero or heroine died of pleurisy, which depressed him immeasurably. My mother would tiptoe into his room and lend him a book she had finished. Every once in a while, Salomone woke up to find my father or Zarifa standing over the bed, peering at him.

Dr. Grossi's diagnosis had left him bewildered. All his life he had been exceptionally healthy. He stood at over six feet, nearly as tall as Oncle Leon. It was incomprehensible that he could feel so sick and beaten down.

It was surely a coincidence that at exactly the same time his parents and older sister were rounded up for deportation, Salomone was felled by this awful malady. He had no idea, that terrible January when his life hung in the balance, that his parents and sister were also prisoners, albeit in a far more sordid kind of jail, in Milan, and confronting a deadlier and more ruthless enemy than pleurisy. Arrested in December as they prepared to flee across the Swiss border, Bahia, his mother, along with his father, Lelio, and sister, Violetta, found themselves among the thousands of Jews who had waited too long to leave their beloved Italy.

NEVER HAD A DOCTOR's orders been taken so literally or implemented so zealously.

Healing Salomone became my grandmother's sole endeavor.

Other women would have felt overwhelmed and resentful at having to care for a desperately ill grandson as well as a pregnant daughter-in-law. But Zarifa, with her profound sense of familial duty, embraced the responsibility.

Her blue eyes gleamed at the challenge.

My grandmother viewed cooking as a kind of black art—part skill and part magic. She was surprisingly lithe in the kitchen, considering her advanced age, able to maneuver from pot to pot, stirring a bit here, adding a spice or condiment there. Then there were the pieces of apricot that she inserted anywhere she could—inside a chicken breast, beneath a steak, alongside a pot of stuffed grape leaves, or in the massive fish from the Nile, the *bouri,* that both her son and grandson loved above all foods. She would use dry apricots because it was so hard to obtain the fresh juicy ones except during the brief apricot season.

She would have liked to impart some of the secrets of her cooking to Edith, to share with her recipes of old Aleppo. Alas, even after she became pregnant, Leon's bride continued to show no interest in Zarifa's bubbling pots and pans.

At 6:00 a.m. each morning, my grandmother would appear by Salomone's bedside. Lightly tapping him on the shoulder, she'd hand him a tray with half a dozen raw eggs. He was always surprised at how hungrily he ate them.

An hour later it was time for breakfast. The rest of the house was awake by then. Leon, already back from synagogue, sat in the dining room enjoying tea with milk, while Edith sipped sweet black coffee. Zarifa would break away to prepare a special tray for her grandson. She loaded it with fresh milk, purchased that morning from the man who came each day with his cow and goat to the back of the house and asked her to choose which milk she preferred. She often chose cow's milk, which was tastier and costlier than the thin, inexpensive, slightly discolored goat's milk. She poured it into a special outsize bowl for her grandson, along with bread and cheese and a tub of fresh butter, then sat down near the bed and watched to make sure he ate.

By ten, the house was silent again. Leon had left to attend to his business, and Edith had returned to her room and her books. Zarifa went back to her grandson's bedside to offer him his mid-morning snack: six bananas.

Shortly before noon, she was back in her kitchen, grilling him a steak. He'd find apricots tucked under the meat, giving it a piquant flavor.

By one, the family gathered again for lunch. In his bedroom, Salomone was offered a plate that overflowed with whatever Zarifa and Edith were eating in the dining room. Typically, that meant rice and vegetables, along with chunks of stewed meat or chicken.

After having consumed the equivalent of four meals, Salomone was allowed to rest. Zarifa gave the maid strict orders not to make up the room as he slept, though only for a couple of hours, because she believed that to defeat the pleurisy, he had to be eating at all times.

By three, she was back at his side, carrying a tray laden with four or five more bananas, which brought the day's total to almost a dozen. The doctor had stressed the importance of calcium, but it wasn't readily available because of the war. Eating twelve bananas a day helped give my cousin the needed vitamins even as my father arranged to buy black-market calcium tablets.

Afternoon tea was a ritual Zarifa relished, though she substituted strong black Turkish coffee for tea. When Salomone became ill, she used the coffee he adored as an inducement for him to eat the fresh cakes and rolls she had purchased.

She was delighted to see how, without prodding, he polished off the milk and coffee, along with the pastries, jam, fresh rolls, and butter.

The family, with the exception of my father, came together again at dinner, which was served at eight o'clock. It was the meal in which Zarifa truly outdid herself. It didn't matter that Salomone had already eaten six times when evening rolled around. My cousin was fully expected to consume multiple courses.

That afternoon, my grandmother would have dispatched the maid to Zamalek, the most upscale neighborhood, to buy a kilo of cherries, which were a great delicacy, and hard to come by. They were needed to make meatballs with sour cherries, a dish that took hours to prepare, and that she had learned in her mother's kitchen in nineteenth-century Aleppo. Zarifa would knead half a dozen spices into the chopped meat—cinnamon, of course, along with salt, black pepper, and *baharat*, a kind of allspice. She'd add tamarind and a large spoonful of sugar into the sauce she made with the stewed cherries, then, remembering how sick her grandson was, and the baby that Leon and Edith were expecting, she'd throw in another spoonful of sugar. The effort was worthwhile; how wonderful to see Salomone devour the dozens of miniature meatballs, which were sweet and tangy at the same time, like nothing he had ever tasted before or would ever taste again.

The high point of dinner was also its finale. Zarifa would arrive bearing her signature dish, an immense platter of rice topped with *mesh-mesh,* juicy apricots that had been cooked for so many hours they had melted into a kind of syrupy marmalade. No matter what else she had made, lamb stew, steak, okra, or chicken, it was considered a must to end the meal with a plate of rice topped by *mesh-mesh.* Salomone gleefully helped himself to plate after plate of apricot-laden rice. He loved whatever his grandmother loved and was convinced, as was she, that the apricots would single-handedly vanquish the pleurisy.

Edith was the only one who showed little enthusiasm.

Zarifa tried to ignore the fact that her daughter-in-law ate only the fluffy grains of white rice, pushing aside the stewed fruit she had so lovingly prepared.

Two months and hundreds of Zarifa's meals later, it was time to return to Dr. Grossi.

When Salomone got up from bed to get dressed, he found that none of his clothes fit. He had gained more than fifty pounds. Leon instructed his nephew to put on a pair of loose cotton pajamas. He took Salomone's arm while the porter ran to summon a taxi.

Salomone looked nervously up and down the street to see if anyone would notice that a grown man was wearing pajamas in broad daylight.

"Seulement les fous s'habillent comme ça," he grumbled; Only crazy people dress like this. But my father ignored him, and the taxi raced through the bustling streets.

Upstairs, Dr. Grossi had to contain a laugh. His "cure" had worked beyond his wildest imaginings. He couldn't find a hint of pleurisy.

MALAKA NAZLI WAS BEGINNING to feel too small to accommodate Leon, his young bride, the baby on the way, my grandmother, and Salomone. My cousin was twenty-two, no longer the skinny, nervous youth that he'd been when he joined the household seven years earlier. He had a good job, friends, even girlfriends, and he knew that he should think of living on his own. Yet even he was stunned when, not long after his recovery, and without even giving him notice, my father informed Salomone that he would have to move out immediately.

My cousin couldn't help wondering if his friendship with Edith was to blame. The two had grown close in the past year. Was Oncle Leon resentful or even jealous?

It was no secret that Edith adored her husband's urbane Milanese nephew, and considered him the only real friend she had in the house. She viewed her mother-in-law with a gimlet eye and found her judgmental and oppressive, despite her kindly airs and solicitous manners.

Edith assumed her life would change once she gave birth. Her husband, who took his familial duties so seriously, would surely recognize the need to stay close to home. She pinned her hopes on the child she was carrying, who would validate her worth, redeem her in the eyes of the mother-in-law who seemed to find her inadequate and the husband who didn't care enough to stay with her through the night.

Her more cynical side knew this was probably wishful thinking, as elusive and intangible as the apricot season, which is so brief and fleeting as to seem illusory.

"Fil mesh-mesh," goes a popular Arabic saying; When the apricot season comes.

What it really means is: Don't bet on it. It will never happen.

On March 6, 1944, the midwife was summoned to Malaka Nazli.

My father, Leon, planned a major celebration to mark the birth of his son. He set out early in the morning to the Congregation of Love and Friendship, bearing a double portion of coffee and sugar and other treats and delicacies.

"Une fille?" he said in disbelief when he returned and the midwife handed him the pretty dark-haired infant.

"Ce n'est pas possible."

He was so disappointed that he left my mother and his newborn daughter and, hailing a taxi, went to the café where a year earlier he had fallen in love with Edith. Seated at his favorite table at the bar, he ordered an arak, and then another, and another. He stayed out all night, unable to hide his dismay, unwilling to face my grandmother, who had wanted a boy almost as much as he. It seemed not to matter that the infant was to be called Zarifa in her honor. That was the way of Old Aleppo, where a father has the privilege of choosing the name of his firstborn.

Years later, after my mother had told her the bitter story of her birth, my sister would steadfastly refuse to be known by that name. Though she was called Suzette from an early age, for that was the way of modern Cairo, where families conferred European names on their youngsters to help ease their way into colonial society, official documents still listed her as Zarifa. From the time she was a young girl, my sister demanded that all traces of her Arabic name be expunged from the records. As she raged and raged at my father, it was as if she were seeking a way to punish him for that original sin, for the fact that he'd had to drown his sorrows over the news of her birth.

CHAPTER 3

The Lost Uncle

I n the mid-1940s, Salomone was enlisted to make a reunion possible between the man whose name he shared and Zarifa. My uncle Salomon, the priest, who had left home as a teenager in 1914, was now pleading to be allowed to visit Malaka Nazli and be reunited with his mother.

My grandmother wasn't well; the news from Italy, the possibility that she had lost another daughter, was almost more than she could bear. Alone in her kitchen, Zarifa was inconsolable.

Could he see her one last time, her son, the apostate, asked?

Père Jean-Marie, as my uncle now called himself, was living in a Benedictine monastery in Jerusalem. He had been in the Holy City since coming to Palestine in 1925, and had enjoyed little contact with the family since leaving home as a teenager, abandoning Zarifa and his nine brothers and sisters to embrace a very different set of Brothers and Sisters.

He knew, of course, that he was considered a pariah. But he still felt that he would be granted a final audience with his mother, especially after the family had reached out to him, seeking help tracking the whereabouts of Salomone's parents and sister. He had received a letter from his namesake asking if the Vatican could find out what became of them after they boarded the cattle train from Milan to Auschwitz. My father himself had urged that the lines of communication be reopened, believing that his brother could use his Vatican connections to solve the mystery. "Il faut faire des enquêtes," my dad would say over and over— One must make inquiries.

In the years since my uncle had left Cairo, a number of myths had proliferated around him and his career within the church. He was said to be high up in the Catholic hierarchy, a monsignor perhaps, even a cardinal—or on the road to becoming one. He was whispered to have close ties to the pope. There were even stories about his heroism during the war, rumors he had helped smuggle dozens of Jewish children who had fled Nazi-occupied Europe into Palestine.

The same family members who professed horror at my uncle's embrace of Christianity were the ones who felt compelled to build up his accomplishments within the church, as if to say, If he had to be a priest, let him at least be a great priest.

In truth, Père Jean-Marie's standing within the Catholic Church was vastly exaggerated. My uncle enjoyed a perfectly respectable career, but he was hardly an intimate of popes and cardinals. He was at most a competent, respected, and—except for his background as a devout Jew—rather ordinary member of his Benedictine order.

But he had spent two years in Rome, so he certainly knew whom to ask about the fate of his own sister.

Was there any chance they had survived? Père Jean-Marie asked his friends and associates in Rome to trace the whereabouts of Bahia, his sister, her husband, Lelio, and their twenty-year-old daughter, Violetta.

The winter of 1945 was a season of wrenching questions and vague, desultory answers. Alas, the Vatican could only confirm what the Red Cross had already gleaned: that the family had been arrested and jailed, then placed on a transport to Auschwitz. After that, all traces of them had vanished. There were no records showing they had been exterminated—and no signs indicating they had survived.

Salomone (standing, right) next to his sister, Violetta, and his parents, his two brothers in the foreground; Italian Riviera, 1937. Violetta, along with his mother and his father, perished at Auschwitz.

Perhaps that was the true horror: there could never be a resolution, a definitive word on their fate, never a death certificate or a burial site.

But Père Jean-Marie had done what he could. Would the family now honor his request?

My father refused to budge. He wouldn't allow the priest to step foot in Malaka Nazli.

In his mind, his older brother had brought nothing but dishonor to a family that prized its good name above all. In the close-knit neighborhood of Ghamra, all the neighbors knew about his apostate brother, and as Leon walked to temple each morning, regal and dignified in his white suit, our neighbors would shake their heads in sorrow that someone as devout as the Captain would be forced to endure such a tragedy.

My grandmother herself was torn. This deeply religious woman still spoke wistfully of the family-owned synagogue in Aleppo, and she had never recovered from the shock of Salomon's conversion. But she also missed him, and she wasn't well, and time was passing.

Would she really die without seeing her son once more?

A compromise was finally reached, brokered through intermediaries, as no one would admit to any direct contact with the priest. Zarifa would meet with my uncle at a prearranged time and location, not far from the house, if two conditions were met:

He would not wear his black priestly habit, and he wouldn't carry a cross.

The day of the reunion, a taxi was summoned to Malaka Nazli. My grandmother, holding on to Salomone, emerged from the house, wrapped in her shiny black *chabara*. Together, my tall, gangly cousin and my frail, petite grandmother made their way to a small house on the grounds of a nearby convent where Père Jean-Marie was waiting.

My uncle was in civilian clothes, as promised. Zarifa burst out crying. Once her most promising child, the one who had caught the eye of all his teachers at the Collège des Frères with his dazzling mind and his facility with most subjects, especially math, where he solved the most complex theorems and equations effortlessly, Salomon would be the one who would help the family recapture its lost greatness. He was destined to go far.

But "far" wasn't supposed to mean total estrangement from all that the family had held dear for hundreds of years. After he had left Cairo, the letters arrived from exotic destinations: Lanzo, Rome, Louvain, Issy, and, finally, Jerusalem. They went unanswered, of course. Still, how odd, Zarifa thought, that her son had ended up in the Holy Land, the place where Jews dream of settling.

Yet the corner of Jerusalem where my uncle lived had a unique history of wooing Jewish converts. The Benedictine monastery in Ratisbon had been founded in the nineteenth century by a French Jew and banking heir named Alphonse de Ratisbonne, who claimed that a miraculous vision of the Virgin Mary had led him to embrace Christianity.

After establishing a Catholic mission in Jerusalem, he devoted his life to doing good works. Beloved by the Vatican, he created an order of nuns, the Sisters of Zion, as well as building the massive stone monastery where my uncle would settle years later.

Salomone watched silently as our uncle tried to comfort Zarifa— priests were supposed to be good at that. He failed, of course, because he couldn't make the one pronouncement that would have effectively wiped away her tears, the one that she and the rest of the family had been waiting for him to make since 1914: that it had all been a mistake, that he had never intended to stray so far, and that he was coming back to Malaka Nazli and returning to the faith of his ancestors.

That was the dream, the fantasy none of us could ever relinquish— that my uncle would abandon the priesthood, beg his family to take him back. And it didn't change from year to year, or decade to decade, or generation to generation. The same hopeless longing of Zarifa was shared by her children, and then her children's children.

Yet there was a softening of attitudes. Marie, the youngest of the ten children, and perhaps the most tenderhearted, was the first to forgive him. My aunt decided to end the schism that had existed for so long. Against the advice of my father—and even her own husband—she opened her house to Père Jean-Marie and allowed him to come meet her own children and as many of the nieces and nephews as she could gather.

Tante Marie was a kindly soul, all softness and curves and compassion, the embodiment of femininity. But she could be every bit as authoritarian as the men in the family. Once she made her decision to receive the priest, no one could dissuade her, not even Leon, the sibling she loved and respected the most, and the only one she feared.

Tante Marie was convinced that the moniker Jean-Marie was a tribute to her. It didn't matter that everyone laughed at her for her foolish notion and said it was only a coincidence—the name Marie was extremely popular among Catholics to honor the Virgin Mary.

But to Tante Marie, it was a way her brother had found to maintain a link to the family when all links had been severed.

Père Jean-Marie arrived at her home dressed in flowing white. He carried beautifully wrapped packages he distributed to the children who gathered around him, delighted by the attention they were receiving from this stranger who looked so familiar, somehow, with his fair skin, aquiline nose, and intense green eyes. Only the beard was jarring; the men in the family tended to be clean-shaven. Still, he was jovial and charming, embracing his nieces and nephews one by one, their very own Jewish Santa Claus.

Reigning over the festivities was Tante Marie, who sat there beaming.

He was her older brother and she loved him and nothing, not even the Church of Rome, would be permitted to erase the bond between them.

MY PARENTS' MARRIAGE ENDURED its first rocky year. The relationship survived both Leon's return to his restless ways and the arrival of a daughter instead of a son. Malaka Nazli was far from joyous. It was a house of tears, steeped in mourning in the wake of the news about Bahia. My mother was struggling to care for my infant sister as well as looking after Zarifa, who was increasingly frail. My grandmother was no longer able to stand for hours at her beloved Primus, and retired to her room.

Barely a year after my sister was born, Mom found herself pregnant again. In May 1946, she gave birth to the longed-for son, my brother César.

At the bris held in Malaka Nazli, Zarifa, summoning one last time her legendary strength, carefully handed the infant over to the mohel, the man who would perform the circumcision on a satin pillow. The mohel dipped his index finger in a cup of wine and gave César three drops, intended to numb the pain.

Several months later, my grandmother died, still grieving over the loss of her daughter but heartened that she had lived to see Leon settled, with a son and heir. His fretful bride and incessant wanderings were almost trivial, incidental details to this indomitable matriarch who had ruled with an iron hand, even when her hand was old and frail. To the end, she had kept her focus on the essentials as defined by Aleppo: faith, honor, and family.

One more tragedy cast a shadow over Malaka Nazli. My father's nephew, Siahou, Tante Leila's son, jumped out the window of his

Edith holding Suzette, the oldest, and Leon holding César, his firstborn son and heir, Cairo, 1946.

mother's house. His suicide was never talked about and never explained.

With Zarifa gone, my other grandmother, Alexandra, became a more frequent visitor to Malaka Nazli. She would arrive every day, and knock rapidly four times, tap tap tap tap. Once inside, she'd settle on a chair and, taking my sister and César in her arms, proceed to rock them and sing to them in Italian. Unlike Zarifa, who only spoke Arabic, Alexandra never spoke Arabic, and she only wandered to the kitchen to retrieve a small cup of steaming *café Turque*.

Alexandra of Alexandria—even more than with Zarifa and her kings, there was an apocryphal quality to the stories about her and her gilded past.

In my mother's telling, Alexandra was a creature both fantastic and fatally flawed. She had lived a life of extraordinary indulgence. The daughter of doting, wealthy parents, Alexandra was lavished with the finest that money could buy. As a young girl, she'd had maids and nannies to attend to her every need.

"Why," my mother, Edith, loved to recall, smiling, "Alexandra couldn't even comb her own hair." Each morning, the governess took it upon herself to brush and braid the young girl's long dark strands of hair, and tie them back with satin ribbons. Perfectly coiffed and dressed, Alexandra would proceed to the parlor for her daily piano lessons. She had the finest instructors in all of Alexandria.

When she was old enough to attend school, her parents enrolled her in a convent run by nuns. Devout Jews, they were also consummate snobs, and no schools enjoyed the cachet of Egypt's Catholic convents. Each morning, a maid would walk Alexandra to school, and then return in the afternoon to take her home.

The nuns were strict and prided themselves on a no-nonsense, disciplinarian brand of education. Alexandra seemed hopelessly weak and vain—someone who needed to be toughened up so she could face the world.

At noon, her parents sent over another servant with a tray and a hot meal, since Alexandra complained she couldn't eat the slop that was served in the common lunchroom, yet which the rest of her classmates seemed to enjoy.

But she couldn't eat what the maid had brought over either. She simply fasted day after day, and her *évanouissements*—fainting spells—made her the talk of the school. Mostly she felt self-conscious over being "different"—a Jew among Catholics, a rich girl among those who were merely well-to-do, a sensitive soul among ruffians. Her sense of isolation only increased, and she plotted to leave the convent at the first opportunity.

That opportunity came in the form of Isaac Matalon, a womanizer with a mysterious income visiting from Cairo. Charmed by the lovely teenager and no doubt finding her an easy mark, he persuaded Alexandra to abandon her parents, her home, her convent, and her city, and accompany him back to Cairo. They left together after a wedding ceremony her parents didn't even attend, and moved into his dingy apartment in a deeply impoverished area of narrow, windy streets and alleyways that seemed worlds away from Alexandria.

The shabby residence helped illuminate the secret of Isaac Matalon's income: he had almost none. He survived through his wits and charm. The year was 1921, Alexandra was eighteen and, within a few months, pregnant. Isaac, a widower with grown children including Oncle Edouard, who would one day accompany Edith and Alexandra on their first visit to Malaka Nazli, was in his forties, possibly even older, since he was as unreliable about his age as about his louche activities.

Isaac was hopeful that Alexandra's parents would bless the union and help out the couple, if only for their daughter's sake. What he didn't foresee was that his in-laws, bitterly disappointed at Alexandra's choice of a husband, would cut her off completely.

The Cairo apartment was so unlike the spacious villa by the sea where she had been raised, and it felt even more cramped after she gave birth to a little girl, my mother, Edith, in 1922, and then, a couple of years later, to a boy, my uncle Félix.

Alexandra felt completely adrift without her maids and her beloved *gouvernante* to help her take care of the house and the children and herself. She didn't have a clue as to how to keep a room clean and tidy, or the faintest notion how to fix a meal for her husband and two infants, as she never learned to cook. She still couldn't even brush her own hair. Without her parents' help, she experienced abject poverty for the first time. There was nothing she could pawn for extra income. Her parents hadn't even allowed her to take her clothes or jewels.

They had relented on a single object—Alexandra's piano. There it stood, stately amid the squalor. Edith and Félix were left to their own devices, neglected and forlorn and, at times, ravenously hungry as Alexandra played and played.

It wasn't that my grandmother was uncaring, my mother would always insist. At her core intensely kind and loving, Alexandra was simply one of these people who couldn't cope on their own, who needed others to get through one day to the next. My grandmother had received a wonderful education in her years at the convent, knew how to converse fluently in Italian, and had a finely honed literary sensibility in addition to her musical skills, but the nuns hadn't prepared her for a life as a wife and mother.

This wasn't what Isaac had bargained for in a wife. Their life together unraveled, and more and more, he left her alone with the children and took off for destinations unknown. Still, those times he was home, he managed to be affectionate to his son and daughter, and Edith, in particular, adored him. If he had harsh words to dispense, they were directed only at his wife. He had come to hate the very qualities that once drew him to Alexandra—her vulnerability, her delicate and intensely fragile nature.

Another child, a boy, came into the world even as my grandparents' marriage was dying. The bonnie baby had soft dark hair, blue eyes, and a gleeful, hopeful disposition. His Hebrew name, after all, was Eliezer, which means "God will help me, God will take care of me." My mother, a little girl of seven or eight, watched over him while her parents sparred, more absorbed in their hatred of each other than in their love for their children. Fights at home grew more bitter, the scenes of recrimination became more frequent.

It all came to an end one morning.

Isaac announced he was taking the infant out for some fresh air. "Edith, *cherie,* could you please dress the baby?" he asked his daughter very sweetly. The child was only a few months old at the time, and couldn't walk or talk. She put him in a fresh cloth diaper—only a cotton diaper, my mother remembered, not even shorts or rompers—and *une flanelle,* a white cotton T-shirt.

The diaper wouldn't stay up, so Isaac offered one of his ties as a makeshift belt. It was wide and red and silk, so incongruously long that it went around the baby twice, at least. Edith gently combed his soft hair and rubbed the popular eau de cologne Arlette on his arms and legs; babies were known to love the refreshing scent.

The baby was gurgling and smiling, clapping his chubby hands. He seemed delighted at all the attention; he played with the tie, trying to undo it.

"On va faire une promenade," Isaac announced, placing the infant in the carriage—We are going for a stroll. My mother kept fussing and fussing over her little brother, not quite willing to let him go, maybe because he was especially sweet and loving that morning. Alexandra kissed the baby as she had done a hundred times, a bit distractedly, absorbed by a novel she was reading.

Isaac returned later without the child or his stroller, and informed his wife what he had done. He had sold their blue-eyed baby boy at the souk, the Arab marketplace.

They simply couldn't afford another mouth to feed. There was no way she could handle another child. It was for Alexandra's own good, he said. Then he turned around and left, never to be part of the household again.

That is when Alexandra began screaming. She screamed so loud that her cries reverberated across the neighborhood. Her sobs were heard in the *kuttabs,* the small, homelike synagogues where men gathered to pray at any hour of the day or night, and echoed across the dusty alleyways and shabby streets, and were heard in the communal baths, where women came to wash away their impurities.

Egypt was a country where professional mourners were often heard crying in the streets. "Someone must have died," strangers thought as they continued walking.

With nothing to fall back on, Alexandra sank into a kind of madness. Edith recalled how day after day, she'd watch her mother from the window as she walked to and from their home, an increasingly thin and haggard figure, old before her time, proud yet desperate, pleading for bits of donations from the Jewish charities that were active in the community.

Whatever money she collected went to care and feed Edith and Félix. She kept almost nothing for herself, and survived on cigarettes and cup after cup of the exceedingly strong *café Turque* that she was constantly preparing in the kitchen. What had happened to her beautiful blue-eyed child? Was God indeed watching over him?

Every few weeks, Alexandra would take Edith to visit some wealthy relative in Cairo—a niece, a cousin, a beneficent uncle—who knew of her financial ordeal and had agreed to help her, however modestly, to pay the rent and buy food for the children. My mother and grandmother would arrive at gated mansions, yet before even going in, Alexandra would remind Mom never to betray how needy they were.

Once inside, maids in uniforms would approach them, offering small sandwiches and petits fours and other delicacies on large silver trays, but my mom, suffering from hunger pangs, would smile and shake her head, *non, merci,* because she was obeying Alexandra, and that meant not letting her relatives know she hadn't had breakfast or lunch. Trays would come and go and Edith, always the model daughter, would follow my grandmother's cue and help herself to one small biscuit, nothing more.

If the visit went well, Alexandra emerged with enough cash to enable them to live a few more weeks, and she'd rave about her wonderful relatives.

With change left over, Alexandra would purchase a ticket at the local movie house. Almost every afternoon, my mother and Félix were left alone in the house while Alexandra, a pack of cigarettes in hand, walked to the movie theater. After the movie was over, she still didn't come home—sometimes not for hours. Instead, Alexandra combed the streets and alleyways, the boulevards and the stalls of the souk in search of traces of my lost uncle.

When she finally returned, she seemed strangely calm, but wouldn't speak to Edith or Félix: her thoughts were only on the child she had lost. She sat at the piano and played. When she was happy and when she was sad, Alexandra played. She played beautifully, my mother insisted, like the great artist she could have been if only she hadn't rushed into marriage with my grandfather Isaac.

Alexandra's search went on for years. She could never stop staring at the infants, the ones in strollers, the ones in cotton diapers who were out enjoying a sunny morning in the company of their parents; she'd examine them too closely. In a corner of my grandmother's anguished mind, her son looked the same as he did the morning he vanished.

Alexandra never forgave her husband. My grandfather finally resurfaced, living in the neighborhood of Daher, minutes away from the family he had left. It was the 1920s; Old Cairo had no tradition of alimony payments or child support. Once they separated—and it is not even clear they were ever legally divorced—Alexandra was on her own, left to fend for herself. The only saving grace was the children he'd produced from his first marriage, Rosée and Edouard. Both were genuinely decent human beings, who felt a kinship and a bond with this woman their father had married and then discarded. Because they were much older than Edith and Félix, they became like parents to them, and for that matter, to Alexandra, since she was so much like a child.

As an old man, my grandfather Isaac succumbed to a massive stroke that left him completely paralyzed, in effect, a quadriplegic. Rosée, his daughter, though struggling to care for several children, took him into her house, forced her sons to share a room with him, and cared for him through his illness. My grandmother and Rosée had remained friends, and Alexandra often stopped by for coffee. Yet she never once went in to check on Isaac, lying helpless in a bedroom in the back. She simply sipped her *café Turque*, chatted a bit with her stepdaughter, then left.

Even decades later, Alexandra's search for my lost uncle continued. Her anguished quest never ended, not even when she was old and bent, and Edith was settled and married with children of her own that would have allowed Alexandra to ease into the far more peaceful role of grand-mother.

Her nervous walks through Cairo simply became more purposeful. She now had a destination after years of having none: Malaka Nazli. She wooed my siblings with her Italian lullabies, and lavished them with all the love and attention she would have bestowed on her stolen child.

My mother's stories about Alexandra always finished with the same refrain. "You, Loulou, are exactly like your grandmother," she would say to me, and she sounded strangely admiring.

As a child, I wondered whether I should take her literally. I pictured myself sitting at a large piano, chain-smoking cigarettes, watching as my house disintegrated, and my husband began to loathe me, and my children wept from hunger and neglect and wandered about in rags. Why, I asked myself, did my mother think I had so much in common with this tormented and demon-ridden grandmother I had never even known?

It was as if she were trying to preserve a piece of Alexandra at any cost, even by remaking me, her daughter, in the image of this woman.

CHAPTER 4

The Last Days of Tarboosh

There is a photograph of the family taken in Alexandria in the early 1950s. My mother is seated at a café table, her hair wavy and lustrous, her legs alluringly crossed. She is the classic Levantine wife—dark, mysterious, slightly melancholy. My father looks smugly toward the camera in his tarboosh, the red fez hat he adores that is the symbol of Egyptian aristocracy. In between them are my sister and my two brothers, César and Isaac, whose births had almost erased the sadness that had settled in Malaka Nazli after the war. My siblings are dressed impeccably, and they are all the very picture of a prosperous Jewish family in Egypt in the early 1950s.

Yet any sense of security was illusory. Life in Egypt had changed drastically since the heady aftermath of El Alamein. The British—bitterly resented by the Arabs for decades—had been barely tolerated during the war, but after the Axis defeat in 1945, there were intensified efforts to force them out of Egypt. As for King Farouk, he had become a symbol of hopeless corruption. The once slender, handsome young monarch, whose reign had begun with such promise,

The family before I was born (left to right):
Edith, Suzette, César, Isaac, and Leon in Alexandria circa 1952.

was now widely reviled for his dissolute lifestyle and grotesque appearance.

Perhaps the greatest irritant of all was the creation of Israel in 1947. The war of 1948 Egypt fought and lost against the new Jewish state continued to rankle. King Farouk was personally blamed for leading his country to defeat, including by the military officers who were actively plotting his demise. And perhaps some of these officers also resented the fact that the king remained close to the Jews of Egypt, who were now seen as the enemy in the same way that Israel was the enemy.

At home, my father tried to hang on to his old ways, but by 1952, that was no longer possible.

Suddenly, Cairo was in flames.

One Saturday afternoon in January, angry crowds rushed through the streets of fashionable downtown, torching all the symbols of luxury and foreign excess—the cinema houses and banks, the private clubs and department stores, the airline offices and outdoor cafés and cabarets that made Cairo one of the most enticing cities on earth, "the Paris of Africa."

Alas, they'd also made the average Cairene feel like a stranger in his own land, because for those who were neither foreign nor rich nor Jewish, much of the city—even a patisserie like Groppi's—was off-limits, and they never felt welcome and most couldn't afford it. Which is perhaps why it, too, became a target of the revolutionary fervor.

As nearly every major establishment associated with the British, the French, or the Jews burned down, Groppi's was also set ablaze. The royal seal was ripped off the front of the restaurant, though not before several of the rabble-rousers escorted the staff, including the chef, the baker, and the Chantilly cream maker, outside to safety. And so the only real victims of the fire were the baking supplies, including several dozen sacks of fine flour and sugar that were carried by the frenzied mob and torched, as if a revolution could be ignited merely by stopping production of scrumptious European pastries and cakes. That night, witnesses would recall, the Cairo air was filled with the heady, troubling scent of burning sugar.

Seven-year-old Suzette was out walking with the maid when they noticed the smoke, thick black swirls of it, in the distance.

Grabbing Suzette by the hand, our *domestique* cried, "Run, run." Together, they sprinted through the streets of Cairo, and for the first time in her life, my older sister felt danger. As they hurried home to Malaka Nazli, they kept turning their heads, unable to resist staring at the clouds of smoke.

Edith had already drawn the shutters. Every once in a while, she'd approach the windows, open them a crack, and scan the horizon. No one knew exactly what was happening—or who was behind it—other than that it was terrifying, and word on the street was that all Cairo was on fire.

Even the king was taken by surprise. Safely tucked away behind the gates of Abdeen Palace, he was enjoying a delicious banquet lunch with hundreds of VIP guests who had gathered to celebrate the birth of his son, Fouad, whose arrival seemed to promise a wonderful year.

The day of the fires, César, six, ventured to the window. Peering through the slats, he could see mobs running down Malaka Nazli carrying lit torches.

"César, éloigne-toi de la fenêtre," my mother said sternly, and tried to steer him away from the window, but he ignored her. My brother was almost in a trance, staring at the crowds running down the street. Would they try to set our neighborhood ablaze? Would they come find us in our home on Malaka Nazli?

It shaped his impression of what came to be known as Black Saturday, or, to the more fanciful, the Day of the Four Hundred Fires, because some four hundred separate buildings were set ablaze. They included one of my father's favorite old haunts, Shepheard's, the world-renowned nineteenth-century hotel that was the most sumptuous emblem of British colonial rule, and his favorite place to conduct business.

How Leon had loved to linger in its oak-paneled bar with the British officers when they owned Cairo and he owned it alongside them.

For days afterward, Jews stayed home, afraid to venture out into the streets, especially downtown. The violence hadn't been aimed at them, of course, not really. It was only the "foreigners" the mobs had targeted, the British in particular. And yet the Jewish community felt intensely vulnerable and feared the worst, wondering if in the eyes of their Arab neighbors, they, too, were now regarded as alien.

The brave hearts who finally wandered over to Suleiman Pasha and other popular areas witnessed scenes of unparalleled devastation that brought to mind Berlin after the war—landmark buildings reduced to ashes, establishments like the Jewish-owned Cicurel department store destroyed, and nearly every major cinema house, from the Metro to the Miami, in ruins. While the marauders had taken care to escort people out of most of the buildings before they set them aflame, there were still several victims. Nearly a dozen Englishmen had perished at the Turf Club, an elite private British club. A young Jewish girl visiting from Alexandria died in the fire at Shepheard's. Looters ransacked stores and banks, grabbing merchandise and cash. By the time the government finally regained control of the city—and Farouk imposed martial law—the devastation seemed less due to unruly mobs than to a strategical and coordinated attack by the king's most powerful enemies.

But it was never clear who these enemies were—there were so many. Fingers were pointed toward the Muslim Brotherhood and the

Communists, elements of the army and some Eastern European embassies and—in the most twisted and bizarre conspiracy theory of all—the British themselves. Who planned the uprising was a mystery that would remain unsolved years and multiple investigations later, though suspicion always fell on the brotherhood and some upstart military officers, or, most likely, a brief, unsavory marriage of the two.

Six months after Black Saturday, on July 26, 1952, King Farouk was forced to relinquish the throne, the victim of a military coup. A few days earlier, a group of a dozen young army officers took control, and while the king tried to call in his chits with various foreign powers, the most that they were willing to do was to make sure he left the country alive. Farouk abandoned Egypt, his palaces, automobiles, and casinos as well as the subjects who had looked to him for protection, like my own family. Fearing for his life, he set sail from Alexandria on the *Mahrousa*. Aboard his elegant yacht, the king was suddenly transformed into that most inelegant of creatures, an exile.

A cortege of crimson Rolls-Royces filed down Malaka Nazli. César watched the royal limousines from the window of my father's room as one after the other, they rolled slowly down the boulevard. They were instantly recognizable, painted in the monarchy's distinctive bloodred. The color red was banned for all other automobiles, even those belonging to pashas and beys. My brother had never seen so many red cars, though who or what they were carrying was unknown: Other members of the royal family? More of the king's vast store of possessions? In happier days, Farouk's Rolls-Royce Phantom had frequently been driven down Malaka Nazli, a main thoroughfare and the most direct route from the royal palace in Koubeh to downtown or Heliopolis. Traffic would invariably stop, as residents and passersby gathered on the sidewalk and stared because the royal convoy was always such a hypnotic sight. *"Ya eesh el malek,"* children would cry excitedly as would some adults; "Long live the king." As he passed, they often caught a peek at him in his tarboosh. The old-timers sadly remembered Farouk in the early years after his coronation, how handsome he looked with his fair skin and gentle smile.

My mother and father stood at the balcony in the dining room, staring at the procession. Dad waved sadly, and both my parents wondered what on earth it all meant, the departure of the red cars, the chauffeurs driving at a funereal pace, so that they seemed to be almost floating along the boulevard, as in a bad dream. Mostly they asked themselves what would happen without the king, who, for all his excesses and foibles and wantonness, had been a good friend to the Jews.

Why, after the birth of his son, Fouad II, a few months earlier, King Farouk had turned to a Jew, Simchon, who was the community's favorite mohel—a man specially trained in the delicate ritual of circumcision. Though there were Muslims equally skilled in the ancient ritual, Simchon was the only man the king would trust to handle the most eagerly awaited baby boy on earth. Egypt's Jews interpreted it as a sign of the monarchy's continued friendship. They had always enjoyed amicable relations with the king, as they had with his father, who'd had a Jewish mistress. The bond had been frayed, of course, after Israel's creation, when Farouk led the Arab countries to war. The exodus of Jews from Egypt had begun in earnest. Even so, the community had a sense of the king as a generally benign figure. For all the major Jewish holidays, he always sent emissaries from the royal court to attend services at the Gates of Heaven. They came in full regalia and sat at the front of the synagogue.

Alas, baby Fouad's birth hadn't reversed the king's fortunes. Perhaps it was because his second wife's name, Nariman, didn't begin with the requisite letter *F*, which was considered a good omen. Some were already whispering that the letter *N* was the culprit—witness the rise of Generals Naguib and Nasser, the two military officers who led the revolt and were poised to take power and overthrow the monarchy.

There was an unbearable sadness to the passing cars. They had led such charmed, self-indulgent lives until now. The king was a passionate collector, who owned dozens of automobiles of different makes and models—the Rolls Phantom, custom-built for him in England in 1940, along with the Ferraris, Bentleys, Alfa-Romeos, and Cadillacs. All were accorded royal treatment, their coats kept buffed and shiny and gleaming.

After the revolution ordinary people turned out in the streets to paint their cars in twenty shades of vermilion. There were suddenly plenty of fiery red automobiles tooling around the streets of Cairo and

Alexandria, as some of the wealthier families went out and bought newly imported cherry-red Packards or Fords of their own.

But the royal fleet, graceful and utterly distinctive, couldn't be copied merely with a coat of paint or a trip to a car dealer. They were unique, an emblem of an era, and my parents knew they would never glimpse them drifting along Malaka Nazli again.

Within days of Farouk's abdication, an edict came down eliminating royal titles. Pashas and beys were no more. The new military rulers wanted nothing even vaguely reminiscent of the royal family, even the red fez hats they loved to wear. They wanted to abolish all vestiges of the family's formidable powers. They proceeded systematically to eradicate any trace of the monarchy, and within a couple of years, even the street names that paid tribute to various royals would be changed.

Our own Malaka Nazli Street, whose name had already been changed when Farouk, in a fit of pique against his mother called it simply "the Street of the Queen," removing any reference to Nazli, was stripped entirely of its name and became the "Street of the Rebirth of Egypt" ("Nahdet Masr"), and then, finally, Ramses Street. Stately Fouad Street, a tribute to the late King Fouad, was renamed "26th of July Street," to commemorate the day the monarchy fell. Farouk Street, rather humble considering its namesake, where you went to buy clay pots and pans for the kitchen, went through different iterations when Nasser and Sadat were in power, and ultimately became "Shariah el-Geish," the Street of the Army.

Any testament to heroes of old also had to be abolished—especially pashas. Glamorous Suleiman Pasha Square, which paid tribute to a French convert to Islam who had built the modern Egyptian army, was swiftly renamed "Talaat Harb," after an industrialist who founded the first bank of Egypt. What was wrong with Suleiman Pasha? King Farouk was his great-great-grandson. Renaming streets had always been popular, but after the revolution it became an obsession. Never easy to navigate, Cairo became a confusing mess of official names and unofficial names, of old names and new names and newer names.

But the edict banishing the tarboosh was perhaps the most telling sign of the ruthlessness of the new order. My father was no longer allowed to wear the red fez he favored above all other hats. To the revolutionary colonels, the cone-shaped hat was a small but potent symbol of the ancien régime. The king had often been photographed wearing the tarboosh, and it was the hat that his pashas had favored. While the tarboosh was popular far beyond the noble court, and even ordinary schoolboys wore the hats as a sign of respect toward their teachers and headmasters, they were linked inextricably to the royal court and a way of life that had to be stamped out.

They disappeared, almost overnight, those dreamy red hats. Suddenly, men began to don caps and felt hats and straw hats and jaunty borsalinos. Ironically, many in the affluent classes took to wearing classic, imported European headgear, though the point of the revolution had been to rid Egypt of its foreign influence. By ruling against the tarboosh, the generals were taking action against an icon whose origin was Turkish but which was now as Egyptian as the Pyramids or the Sphinx. And sure enough, years after the fez was declared illegal, it became increasingly popular for doormen and attendants working the posh hotels around Cairo to wear them as they greeted tourists who wanted a taste of the "real" Egypt.

Leon took his collection of tarbooshes with black tassels and stashed them away in his closet, next to his treasured British regulation pith helmet. He never wore a tarboosh out on the street again, though he was determined to hold on to them, and enjoyed simply fingering their soft velvety contours and playing with the tassels.

EDITH WAS EXPECTING AROUND the time of the abdication. It was a difficult pregnancy, and she wasn't feeling at all well. She was weak and feverish and in pain, yet the family doctor who came to examine her didn't have a clue what was wrong. She grew worse in the final weeks of her pregnancy, complaining of headaches and stomachaches, but everyone assumed it was the child she was carrying, her fourth, that was taking a toll—*Pauvre Edith était si delicate* ("Poor Edith, so fragile")—so no alarm bells were sounded and no special measures were taken.

For years, she'd been helped by Simcha Allegra, the midwife who had delivered all my siblings without any complications. The midwife was so seasoned, so capable, it was almost as if a physician was beside the point.

Except it was different, this time. There were only complications.

My mother was so ill when she went into labor, it was a miracle she survived and the baby was born healthy. Seeing the little girl who was so perfect, so pretty, made everyone feel almost hopeful at first, as if the worst was over. The infant cried on cue and her limbs were well formed and her breathing regular, but most striking were her eyes—a doll's eyes, vivid, dazzling, baby blue.

My brother and sister who had waited anxiously in the dining room throughout the labor were allowed to see the infant, if briefly. Suzette took one peek at the child and declared her beautiful. That was the verdict of everyone who had a glimpse of the newborn—*qu'elle était simplement ravissante*—a gorgeous, magical child. She had a fair complexion and a soft fuzz of light brown, almost golden hair.

They named the baby Alexandra, after my maternal grandmother who was always in our house, fussing over the little girl, who filled her with a joy she hadn't felt in years.

But Edith was getting worse. She seemed confused and delirious at times: she'd wake up to feed the baby and then fall back asleep, exhausted. But it was never a restful sleep. She was always moaning, or crying out for people from the past—her brother, the lost baby of the souk, or else Isaac, the father who had abandoned her.

Alexandra would chain-smoke her cigarettes, wander outside, and return to find her daughter so sick she couldn't open her eyes. It didn't help to appeal to Leon. Alexandra knew from the start that the marriage was an unhappy one, and she blamed my father for Edith's misery and herself for having approved the union that afternoon at La Parisiana. Yet she'd always been afraid to stand up to him or protect her daughter.

Even now, at this critical juncture, with two lives hanging in the balance, my father and my grandmother could barely communicate. Aware of the danger his wife and daughter were in, Leon had summoned all the doctors he knew to Malaka Nazli; yet none seemed to know what was wrong.

Alexandra suddenly discovered a well of strength within her that spurred her to take action. She left Malaka Nazli and walked, cigarette in hand, to the home of her stepdaughter, Rosée. Over a steaming cup of Turkish coffee, she confided the drama that was unfolding in our house.

"Edith est en danger. Je suis folle d'inquiétude," she told Rosée— Edith is in danger, I am worried sick.

In times of crisis, the two women agreed, there was only one person to turn to, Rosée's brother. Oncle Edouard, my grandfather Isaac's son from his first marriage, was a charismatic figure, the head of my mother's side of the family. He had managed to climb out of poverty, and as he prospered, he kept the extended family afloat. Because of his work as a pharmaceutical salesman, he knew all the leading specialists of Cairo as well as the most up-to-date drugs. My mother adored him, this half brother who was more like a father to her.

Oncle Edouard.

Informed that Edith was in danger of dying, Oncle Edouard rushed to summon one of the leading infectious disease specialists in Egypt. Together, the two men made their way to Malaka Nazli. My father greeted them with relief: he may have been the unchallenged ruler of Malaka Nazli, but he always knew when to cede power.

The white-coated physician instantly made a diagnosis. "C'est la fièvre typhoïde," he declared, confident and grim, and everyone felt shaken at his words. Typhoid fever was the scourge of Egypt.

It seemed so obvious now. But while widespread, it was often misdiagnosed. Still, so much could have been done differently, if only they had known. There were medicines that could have been administered, doctors who could have assisted in the delivery, not merely a midwife. Most important, mother and child would have been separated.

The doctor insisted on immediately removing Baby Alexandra from Edith, who had continued to nurse her despite her own debilitated state. But it was undoubtedly too late, he warned somberly. Because my mother had held her and fed her, it was likely the bacteria had spread to the infant.

Malaka Nazli, which should have been a house of joy, was again a house of tears.

Oncle Edouard directed his anger against my father, demanding to know how he allowed his wife to remain in such a state, when he could certainly have afforded to bring in the finest specialists much earlier. Leon didn't even try to defend himself. He was mute, in shock at all the conflicting events—a newborn daughter, a desperately ill wife, news that both were in peril, and all from as common a disease as typhoid fever, which any competent physician should have been able to diagnose.

He had to act swiftly to safeguard the other children. As panic swept the house, both César and Suzette were bundled up and taken by taxi to Tante Marie, while arrangements could be made to protect Isaac, who was only a toddler.

The days turned into weeks. It was such a strange period, when Suzette and César heard nothing about how their mother and their new sister were doing. Because it was the summer, there was no school that would have given their days a structure. My father would stop by to look in on them almost every day, but he was vague and evasive about goings-on on Malaka Nazli.

Alexandra, undaunted by the prospect of being exposed to the fever, would sing to Baby Alexandra, hoping to get her to flutter her eyes open. After washing her hands in the basin filled with disinfectant, she would gently stroke her and hold her close, this child of her heart, the most extraordinary of all the babies Edith had produced, a blue-eyed princess in a kingdom of brown eyes.

But the fever took over and Baby Alexandra's luminous eyes dimmed and her body grew impossibly flushed and warm, and she couldn't breathe.

There was nothing anyone could do.

My mother wasn't allowed to hold her or comfort her, and she was too sick—mercifully, perhaps—to realize that her baby was dying in the next room.

There was a chance that Edith would also not pull through, that she wouldn't be able to overcome the fever. My father opened the house to her side of the family, as they launched an unprecedented offensive to save both mother and child.

After Oncle Edouard had brought the doctor, Tante Rosée all but moved in. She tended to Mom night and day—never going to sleep, watchful for the slightest change, attentive to my mother's every need.

Under the determined care of the indomitable Tante Rosée, my grandmother hovering close by, Edith survived. Baby Alexandra did not.

Whenever my mother asked, "Où est la petite?" she was told that she was too sick to see her daughter, that it would only put the child at risk. Edith would nod and go back to sleep, retreating to near oblivion. She was still delusional, still confused.

At last, my siblings returned from their exile. They went from room to room but found no trace of the child. Malaka Nazli was seemingly the same as before, except that it was scrubbed clean and still bore the scent of strong disinfectant. Everyone was silent, especially our mother, who remained in bed all day.

No one would explain exactly what had happened, no one would say where the baby had gone, and they knew instinctively not to ask questions. It was the way of Old Aleppo never to expose children even to the hint of death. The young were barred from going to a house of mourn-ing. They couldn't pay their respects, or go to the cemetery, or even wear black because that would bring bad luck.

SHE LASTED EXACTLY EIGHT days, Baby Alexandra, though years and decades later, she was still there, deep in our consciousness. No pic-tures of her exist. The tradition was to take a new baby to a professional for a formal studio shot within weeks of her arrival, and that, of course, had not been possible.

César would speak to me of this lovely child with clear eyes, and Suzette remembered the delicate mound of light brown hair that she had briefly seen. We all obsessed about her in our own way, imbuing her with all the qualities we valued. She haunted me, this child who never grew up.

"Loulou, you, of course, are Alexandra," my mother would say to me again and again. For a long time I thought she was comparing me to my grandmother. Only later, much later, did I begin to comprehend that she was invoking the child she missed. I was, to her, the little girl blue, returned to this earth.

MY MOTHER LOST HER beauty. It was a gradual process, which had begun even before Baby Alexandra, but accelerated with the child's passing. Her lovely white teeth had inexplicably started to crumble after César was born. With the years, she was left almost toothless, and it became almost painful to look at her, this young woman in her twen-ties with exceedingly fine features, and the puckered-in mouth of an old lady.

She became a recluse.

Edith rarely left 281-Malaka Nazli. The ostensible explanation for her decline, or at least the one my family embraced, was that Mom, who was anemic and ate very poorly, had suffered a loss of calcium after her second pregnancy, a process that intensified with subsequent births.

She never drank so much as a glass of the fresh, delicious milk that continued to come by way of the goats and cows who still showed up every morning at the back window of Malaka Nazli, with their caretaker offering to fill up a pitcher for only a few piasters. Edith's beverage of choice was coffee—*café Turque*—strong, black, with extra sugar to camouflage the bitterness. When she was done, she would turn the cup upside down and read her fortune in the coffee grounds.

But the absence of joy played as much of a role in Mom's decline as the absence of calcium. With the death of Baby Alexandra, my mother sank into a profound melancholy. The sadder she became, the more she neglected herself. It didn't help, of course, that she had never been to a proper dentist, though there were several European-trained specialists practicing in Cairo in the 1940s and '50s. When she was growing up, she was too poor to see a dentist. After she married my father, she was simply too frightened.

If Edith's prettiness had drawn my father to her, it was almost as if she were thwarting him now. In her anger, she seemed to be saying: "If you will not live up to your end of the bargain, I will not live up to mine."

My father's reaction to the death of his daughter was entirely different. After a formal mourning period, he resumed his nocturnal routine. He went alone and never took his wife along, if only to console her, to pull her out of the thick shell into which she had withdrawn.

For Edith, this was yet more proof of her husband's selfishness, the fact that he couldn't relinquish his pleasure-seeking even after a tragedy. Despite years of marriage, my mother had never understood Leon. My father's habit of leaving home wasn't much different than Edith's retreat into the bedroom they no longer shared. Much as she coped by escaping from the world, he found comfort venturing into the world, finding forgetfulness in the endless nights of poker and dancing and solace in the company of other men and other women.

CHAPTER 5

The Prisoner of
Malaka Nazli Street

My aunts rushed over to our house the instant they heard the news: Edith wanted a divorce. She was threatening to leave with the children, and Tante Marie, Tante Rebekah, and Tante Leila were beside themselves. It was simply unthinkable. This was a family that, in more than half a century in Egypt, had experienced madness, suicide, murder, adultery, apostasy, and the Holocaust. But divorce? Never.

Well, maybe not exactly "never."

There were the failed marriages of Oncle Joseph's teenage daughters, unions that had both ended in disaster. But they came from Zamalek, one of the richest and most elegant parts of Cairo. Maybe they tolerated failed marriages amid the mansions of Zamalek, but here in Ghamra, it was almost unknown.

My aunts stood in the kitchen with my mom, trying to talk some sense into her, but also lending a sympathetic ear. Edith's mourning may have been muted, almost secretive, but her anger was outsize. She blamed Dad—stubbornly, obsessively, and irrationally—for the events surrounding Baby Alexandra's passing and her own brush with death.

"Chez nous, on n'a pas le divorce," Tante Marie declared, while the others nodded gravely.

Leon's sisters were kind, caring women, even oddly modern in their own way, though it was true that Tante Rebekah's thick coils of hair suggested a woman of another century, the way she wrapped them so tightly around her head. They were confident they could navigate the old world of Aleppo that my father stood for, where a woman's desires counted almost for nothing, and the slightly more progressive world of a Cairene housewife in the 1950s.

To their credit, they didn't automatically take my father's side. That's what made them such wonderful women. They were deeply earnest and well-meaning, and my mother felt she could trust them. Though passionate about the need for families to stick together against the world—they had been brought up to value blood relations, above all— they also felt a genuine kinship with *La pauvre* Edith.

Tante Rebekah
with her husband,
1920s Cairo.

Their own households, curiously, in no way resembled ours. Rebekah, Marie, and Leila all had husbands who venerated them and were slavishly devoted. Even Henriette, who was married to Dad's older brother, Oncle Raphael, had put a stop to her husband's fondness for nightly poker games, insisting he stay at home with her and the children; she gave him no choice but to obey. Some women had such forceful personalities they were able to impose their will, even in the male-centric Levant.

It was fine among their Arab neighbors for a man to leave his wife behind and go out for his night of fun and games, to lead, in effect, two lives: they could still command respect at home as a father and husband, yet take off without a thought to be with their mistresses, or their friends, or both.

But not the Jews. Like the Europeans they befriended in the glittering social world of 1940s and 1950s Egypt, if they went out, they went as couples. If it was a night of dancing, the men danced with their wives. If they watched the famed belly dancers who performed around Cairo at the stroke of midnight, their wives sat beside them and laughed and applauded as heartily as they.

Of course, the Jews of Aleppo were a breed apart—intensely Jewish, intensely Arab. Wherever they lived, in Cairo or Paris, Geneva, São Paulo, or New York, their mentality was similar to that of their Muslim neighbors of old, and they seemed to float through the twentieth century oblivious to the social changes, and specifically to the evolving status of women. My father had left Aleppo as an infant, yet it was his defining identity.

Like other Syrian Jews, he played by an ancient set of rules, in which women were adornments—passive, inconsequential, devoid of any power.

Tante Marie, who was closer to my father than anyone else, was troubled by how her cherished older brother behaved toward his bride. She saw an arrogant, cold side of him she hadn't noticed before, or perhaps she had simply preferred not to notice it. She would watch him order Edith around. Later, at home, she complained to her children that he treated his wife not much better than the maid. Her own husband, in contrast, was extremely deferential. He felt privileged simply to have married into so grand a family.

That was the glue holding all these marriages together—that we were descendants of nobility, that we had once dined with kings.

The grandee past was all that mattered, and no one had perpetuated the myth of our illustrious pedigree better than my grandmother Zarifa.

Certainly, Leon approached the world from some godlike elevation, and this was partly due to his extraordinary height and bearing, the fact that he was a *bel homme* and knew it, but also because he had so thoroughly absorbed the fable of our family as told by his mother. He became a living, walking embodiment of all the legends she loved to recount—the private synagogue the family had owned, the thousands of devoted followers who came to pray and study there, the generations of rabbinical leaders and thinkers and scholars, all bearing our name, who had wielded influence and authority far beyond Aleppo.

It is hard for gods to come down to earth, let alone wake up and find themselves married to ordinary mortals. And that, of course, is what my lovely young mother had turned out to be. The delicate, exquisite porcelain figurine he had spotted years back at La Parisiana had shown herself to be a seething bundle of resentments. She had never accepted him as he was, a man who couldn't possibly be chained down to a single house, or a single woman, even for a single night.

Her rival proved more than she could handle: it was none other than Cairo herself, the city whose charms were so boundless that a young housewife, even a pretty and educated one, couldn't possibly compete.

My siblings became hapless witnesses to the constant clashes. The arguments could erupt at any time, early in the morning when Dad came home from synagogue, midday, as he sat down to have lunch at home, hoping to rest during the afternoon siesta, or even as night fell, and they were getting ready for bed, and a soft breeze was blowing. He was getting dressed because his evening was only beginning.

The scenes had been going on since the first year of the marriage, with occasional truces, when the children were infants and distracted her. Now, since the death of the baby, Edith was in a state of perpetual fury, and her bitterness and despair had finally reached his sisters' ears. There were no secrets in Ghamra, and they warned Leon that the threats seemed real this time: Edith was determined to end the marriage.

As Suzette watched in a corner, our aunts calmed Edith down by repeating what women have told each other through the ages. They offered, in effect, a Levantine version of "boys will be boys," telling her *les hommes sont comme ça* and appealing to her common sense. Edith, *chérie*, remember the wonderful family you have. Edith, *chérie*, keep in mind the primary victims of any separation, and besides, what is the harm in a man going out for some rounds of poker every night? Let him live, they'd say: a man has to live. You know, Edith *chérie*, he is devoted to his family, and you are the only woman he loves.

My aunts were too delicate, too sensitive, to remind her that she really didn't have a choice. Cairo in the 1950s didn't look kindly on a woman without a man. Though Edith had been talented enough to land a prestigious job at fifteen as a schoolteacher, it was far from clear if she could work again in these turbulent postrevolutionary times, or that the pay would be enough to support her and the children—that is, if my father even let her keep my brothers and sister. Even in the best of times, tradition, religion, culture, and the law were all on the man's side. But now, with Jews losing their businesses and their jobs because of the ruthless new regime, and the constant sense of danger as the Jewish community unraveled and more and more families left Egypt, a woman alone wouldn't have a chance of surviving.

Besides, my mother still had far too vivid memories of her own fatherless upbringing, when Alexandra couldn't even scrape together enough to feed them, the months and years when she went hungry because they were literally too poor to eat. That is what divorce meant for a woman with no means and why she was, in effect, a prisoner of Malaka Nazli and had been since the age of twenty.

My father, for his part, cast a cool glance on the growing turbulence at home. He knew that he was in a position of strength, and no matter how much Edith wanted it, she would never be able to go through with a divorce. That could only come from him, and he had no desire to get out of the marriage, and no one else he wanted to marry. He had the beautiful children he always wanted, including the sons who would carry on his name. He had fulfilled his obligations. As long as he was home Friday night for the Sabbath dinner—and he always was—then he felt free to go about as he wished the rest of the week.

Through all of Edith's outbursts, it never occurred to my father to take the one step that would have instantly quieted her and reduced tensions: to change his lifestyle, to refrain from going out, or merely to come home a little earlier. Even if he'd wanted to, even for the sake of a fragile and illusory peace, the promise of harmony at home, it would have been impossible.

My father's life at night, his wanderings in white across a darkened Cairo, were as essential to him as oxygen.

FROM AN EARLY AGE, my oldest brother learned to be a buffer between my warring parents. At first, the simple fact that César was the firstborn son, his place secure in a family that valued male children above all others, made him my mother's natural defender and protector. Dad could no longer express annoyance with her because she had produced the desired heir. The fact that César was a quiet, sweet-tempered child made it that much easier for him to assume the role of peacekeeper. He also projected uncanny maturity and understanding. Both my parents felt they could trust him—and each seemed convinced he loved them more.

"Il faut toujours consulter César," my mother began to say, and— voilà—a legend was born:

If you need counsel, even on the thorniest problems, go immediately to see César.

Dad began taking my brother along to work and business meetings. He was only a little boy, seven or eight years old, but my father figured it was never too early to start grooming him to be a businessman and entrepreneur, and lectured him on the need to look and dress impeccably when seeing clients.

Meanwhile, my mother took to confiding in my brother because she valued his quiet insights. To both my parents, he was far more soothing company than Suzette, who was, from the start, such a wayward child. My sister reacted to the troubles at home by behaving ever more fretfully and becoming more rebellious. Early on, my mother had made the mistake of telling her the story of how Leon had reacted to her birth, forever turning daughter against father.

Unlike Suzette, my brother seemed to grow calmer with each of my parents' outbursts. Yet no one was more startled than he when my parents told him they were going on a trip to the Abbassiyah district, and that he was coming along. As he sat in the taxi between Mom and Dad, the ten- or twenty-minute ride to the neighborhood on the far side of Malaka Nazli seemed endless and somewhat scary, since my brother had no idea where he was being taken, and why. He noticed that my parents weren't speaking to each other, that they only addressed him. Mom seemed especially nervous, and she was holding his hand tightly. Dad was cooler and more detached, gazing quietly out the window as the scenery changed from the lively elegance of our boulevard to the poorer, bustling streets of an older Cairo, crowded with open-air stalls where traders and merchants peddled cheap wares.

Occasionally my father would reach into his pocket and offer him a bonbon, which made my brother oddly more anxious, since he wondered what kind of journey they were undertaking; if it was simply a family outing, then why didn't they include Suzette and Isaac, who had stayed behind on Malaka Nazli?

At last, the taxi came to a stop in front of a small building where Cairo's Jewish community maintained some offices. César, still holding both my parents' hands, followed them into the office of a tall, forbidding-looking man who wore a dark skullcap on his head, and whom Mom immediately addressed as "Hacham," the honorific for rabbi. He invited them all to sit down around his desk.

"Quel gentil petit garçon," the rabbi said, praising César as a sweet child. Both my parents smiled graciously, but Mom was clearly on edge.

This was as far as the small talk went. The rabbi immediately began to quiz my parents on the state of their marriage. What was so wrong? Why had life together become so impossible? All the while he spoke, he kept peering at my brother, sitting quietly in the seat between Mom and Dad.

The rabbi noticed that unlike other children he'd observed in similar situations, César had a remarkable ability to sit absolutely still.

"Ce garçon est très bien élevé," the rabbi told my mother, who loved when someone noticed César's perfect manners.

"Votre fils est très beau," he told my father—Your son is very hand-some. Dad nodded, pleased at hearing yet again that he had produced a beautiful son.

With these two deft compliments, the holy man managed to place my brother on center stage—precisely where he wanted him to be.

In Cairo of the 1950s, there were no high-powered divorce lawyers or marriage therapists, the weapons of choice in America for a feuding couple. If problems erupted, members of the extended family—aunts, uncles, cousins, in-laws—would immediately swarm around the trou-bled household, prepared to say and do what was necessary to keep a couple together.

But when that failed, or if it worked only for a time, the rabbis were summoned.

The *hachamim* enjoyed extraordinary cachet in a community that was both devout and worldly. Men and women, rich and poor, the edu-cated classes who lived *en ville* and the impoverished, at times illiterate Jews of the Old Ghetto, and all those in between, were taught at an early age to revere their rabbis, and, most importantly, to defer to their judgments.

Called in to mediate when all else was failing, they were on the front lines in the battle to hold relationships together in the face of hostility, infidelity, abuse, violence, or simply a waning of love and desire—issues that in modern Western societies would lead to a rapid dissolu-tion of the marital contract.

But not in Cairo.

Divorce was extremely rare, and the rabbis, in particular, exerted such a formidable influence on the mores and lives of Egyptian Jewry that they usually prevailed, so that in the unlikely occasion a couple separated, it was the talk of the community for years to come. And while it was certainly easier for members of the wealthier classes to divorce, it was still taboo. It invariably cost one side of the family a fortune—typically, the wife's parents, because in a religious divorce under Jewish law, the husband had all the power. It was up to him to grant his wife a formal writ of separation, a document known as a *get*. If he balked—as many men did, or else made impossible demands for cash, assets, and more—the woman found herself unable to remarry, unable to resume any kind of normal life. She could be rich, lovely, devout—she was still untouchable by other men, a pariah in the community.

It was easier in Zamalek, but deceptively so. The woman who left her husband to return to her parents' villa didn't starve, as my grandmother Alexandra had, but she was still socially ruined. My cousin Marcelle, married off at fourteen to a man in his fifties, begged and pleaded for a divorce from the man she hated, to no avail. She left him anyway. While she was young and extraordinarily beautiful at the time of her separation, no Jewish man would go near her, and she found herself at eighteen facing a life of spinsterhood. Her only options were to marry a Christian or a Muslim. Marcelle opted for a wealthy Muslim and converted to Islam. Her sixteen-year-old sister, Yvonne, fared no better when she left her philandering husband. Her father was able to pay a king's ransom for a religious divorce, but even that wasn't enough. To obtain the *get,* she was forced to give up her baby daughter to her husband, in a deal negotiated by the chief rabbi of Egypt.

Only then would her unfaithful husband agree to free her.

It was no accident that César found himself dragged to this 1950s equivalent of a mediation session. The rabbis' most potent weapons were the children, and they weren't shy about enlisting them as allies, as a reminder to the couple of the high stakes. It was excruciating and even traumatic for the child, but necessary if the family was going to stay together.

That morning, both my parents were offered the opportunity to have their say, but to their astonishment, the rabbi kept turning to my brother.

Who would he prefer to live with in the event of a separation, Mom or Dad? Which parent did he love more?

Because César couldn't possibly have answered, because the entire journey, from the endless taxi ride to the session with this holy man, had the feel of a nightmare, because he wanted to cry, or leave the room, the rabbi was able to make his point: that it was wrong, or worse, *haram*—a sin—to make a child suffer because of a marital dispute, and forcing him to choose between his mother and father was an abomination.

Years later, César would find he had forgotten many of the detailed exchanges that had taken place, but what he could never, ever banish was the sense of hopelessness he'd felt seated in between my parents in the backseat of the cab, and then in the rabbi's office, the sense of being a captive, a prisoner, forced to go along on a journey that could only end disastrously.

"Loulou, never marry a Syrian," my mother would tell me over and over. And with that warning, she offered me a window into her own universe of sorrow, the anger that she still harbored over her years as the prisoner of Malaka Nazli, long after she had been set free from our pretty house with the multiple balconies, and my father was too old and frail and infirm to be an especially fearsome warden.

What drove my mother as she repeated those words to me over and over, as if hoping to hypnotize me, was the profound sense of regret she still felt at having stayed, the fact that she hadn't broken out of 281-Malaka Nazli when she still could, hadn't ignored the tender, solicitous advice of Tante Rebekah, Tante Marie, and Tante Leila, hadn't walked out on my father and the rabbi of Abbassiyah.

CHAPTER 6

The Essence of a Name

Your name is your destiny. Change your name, the mystics say, and you will avert even the most terrible fate.

When I was born, nearly four years after the death of Baby Alexandra, my family felt an overwhelming sense of relief followed by a rush of panic. Simcha Allegra, the midwife who delivered me in the back bedroom of Malaka Nazli, and whose Hebrew and Italian names both meant "joy," stepped into the parlor to tell my father the good news. He had a new baby girl, the midwife announced, and both mother and daughter were well. She spoke crisply and with authority, as befits a woman who has delivered hundreds of infants.

It was one of those rare times in this vast Syrian clan obsessed with producing sons that a baby girl found herself *wanted*. My birth was seen as a sign that the family's luck was changing. The all-merciful God who had taken away one little girl from Malaka Nazli had now shown his infinite compassion by giving back a little girl to Malaka Nazli.

But with the elation came fear that turned into paralysis. Everyone

seemed stumped as to what to name me. Names were critically important in determining a person's fortunes, yet because of the shadow cast
by the death of my sister, deciding what to call me turned into a high-
stakes game of chance. No one wanted to play.

My mom secretly thought of naming me Alexandra. She was so
happy to have given birth to a little girl, maybe even that girl, minus the
blue eyes, she thought to herself, because she was superstitious and
believed in the return of the dead. If I were really Baby Alexandra,
come back from the other world, then wouldn't it make sense to give
me her name? It was so tempting, so human, to want to try again—to
break the spell and defy the odds and give life to a human being who
would have that name yet would thrive and prosper and not fall prey to
the evil eye. My mom was convinced that anyone who was too beautiful
or too good risked being destroyed *par le mauvais oeil.*

But who would impose such a burden on a newborn? The name was
surely cursed. Look what had happened to her daughter. Look what
had happened to her mother—abandoned and bereft as a wife, a beg-
gar living off the alms relatives deigned to give her, reduced to smoking
cigarette after cigarette, and running off each day to the Cinema Rialto
to escape her woes.

My grandmother would wander into our house on Malaka Nazli
every afternoon as was her wont, the mane of soft dark hair that a governess had once lovingly brushed and tied back with satin ribbons now
gathered carelessly in a chignon at the back of her neck. Her fine chiseled features and smooth skin, golden brown from constant walks under
the hot sun of Cairo, made her face look years younger than she was,
although her posture, the fact that she had become almost hunchbacked, made her appear many years older.

She was such a loving creature, *la Nonna,* so tender despite her hard
life, so giving despite the fact she had nothing to give.

Though penniless, Alexandra would come laden with gifts. Typically,
these were stacks of books and discarded issues of French magazines
she received from the phalanx of relatives who looked after her.

After I was born, she arrived with packages of impossible splendor.
Alexandra regarded my birth as almost a religious occurrence. She had
adored the blue-eyed girl who had briefly borne her name, had mourned

her passing with nearly the same intensity as my mom, and had been plunged into the same boundless despair. But it was her own child she missed, of course, and was still mourning so many years later. For Alexandra, every pathway led back to the lost boy of the souk.

My siblings and I were the beneficiaries of her thwarted love. In my case, she prevailed on some wealthy cousins—members of the Dana clan who lived in Cairo—to give her the baby wear their own children had outgrown. She proceeded to lavish me with blankets and hand-embroidered bibs and soft towels and impossibly delicate white voile dresses fit for a child princess so that even my father, a connoisseur of fine fabrics, was impressed. My grandmother saw to it that I was covered with all the lace and linens and soft cottons a fretful infant could want.

Mostly, it was her voice that soothed me when I couldn't stop crying.

There was no piano for her to play anymore. The single room she occupied not far from Malaka Nazli had only a cot and a burner on which she kept the coffeepot where she prepared her ten and twelve cups of dark, sweet *café Turque*, never bothering to rinse it.

Without a piano, Alexandra's musical talents found an outlet in the continental melodies she sang to my siblings and now to me. They were songs from the war years—World War I as well as World War II—and she sang them in her lovely voice in both French and Italian, and the family was hypnotized by this music from a lost world.

My grandmother would shyly approach my crib and begin to sing, not lullabies but grown-up tunes about men who left and the women who pined after them, all set in picturesque lands where the houses were white and the gardens bloomed, the sea was blue and love was forever—"Plaisirs d'Amour," "J'Attendrais," "Santa Lucia," and her most favorite of all, "Torna a Sorrento," Come Back to Sorrento.

"Guarda, guarda questo giardino. Senti, senti questi fiori d'arancio," Alexandra sang, a small angelic figure hovering over my crib—"Look, look at this garden, inhale, inhale the orange flowers." She sang to me years before I would learn to appreciate Italian gardens, and the heady scent of orange flowers, and the Sea of Sorrento and my hopelessly romantic and ill-fated grandmother, winding her way through the streets of old Cairo.

I LINGERED FOR DAYS—a child without a name.

Legend had it that hundreds of years back, my ancestor, Rabbi Laniado of Aleppo, stricken with a fatal malady, had glimpsed the Angel of Death lurking by his bedside. Since the doctors were powerless to save him, he took matters into his own hands—he changed his name and instructed his family to proclaim he was no longer Rabbi Laniado. The stratagem worked like a charm. He tricked the Angel of Death into leaving his room, and survived to a ripe old age.

Yes, a name could do that—it could quite literally mean the difference between life and death.

As they pondered and debated, my family passed on a host of perfectly fine choices. My sister's refusal to be known as Zarifa had put the fear of God into my parents about Arabic names. This meant that my legion of aunts and cousins, any of whose names could have been worthy choices—Bahia, Ensol, and my favorite, Leila, which meant "night"—weren't considered. My father, recalling the tearful scenes and wrenching arguments with my sister, who insisted on being called Suzette, didn't push any of them.

Yet tradition dictated that I be named after a relative. My sister was still identified as Zarifa on official documents. César was Ezra, after my paternal grandfather, who had died decades earlier. My other brother was named Isaac, after my maternal grandfather, who had betrayed and abandoned the family, yet was so beloved by my mom she chose to ignore the inherent risks.

Finally, César broke the impasse. He had a crush on a teacher at school, Mademoiselle Lucette, and when he told her of my birth, she embraced him and gave him a kiss. Flush from the show of affection, César ran home and told my startled family that I should be named after his pretty Parisian *institutrice*. My parents looked at each other, then at my sister and at Nonna Alexandra, lingering by my crib, softly crooning her Italian ballads.

My father nodded yes, and my mother lit up, and my sister offered no objections, and Alexandra continued singing, and that was that.

"Ça lui portera bonheur, Inshallah," my mother remarked; It will bring her good luck.

BARELY A MONTH LATER, the world seemed about to explode: on October 29, 1956, war broke out in Suez. I screamed when planes flew overhead and air-raid sirens warned of an imminent attack. My mother said that my cries were so shrill and piercing, they left everyone in the house feeling rattled and unsure about what to fear most, another incursion by the British, French, and Israelis, or another outburst from Loulou.

That was how I was known—Loulou. It was so much more than a nickname, it was also my persona, the youngest of several siblings, doted on and feted by all. My father deftly took my Parisian moniker and made it sound Arabic simply by emphasizing the first syllable—LOUlou, not LouLOU, as the French would have said it.

I was a frost blossom, the infant daughter of a man in his late fifties.

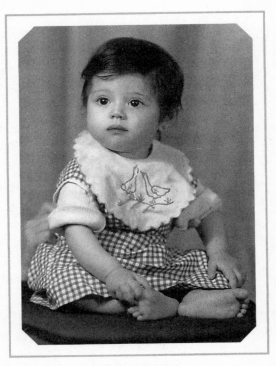

Loulou as an infant in Cairo.

I came at a time in his life when he was feeling discouraged, and the world seemed to have lost much of its promise and possibilities, so that he greeted my arrival in the way a patch of winter shrubbery embraces a small flower that manages to sprout in its midst. He was still shaken by my mother's demands for a divorce, her refusal to tolerate the freedom he found so necessary to survive, and befuddled by all the disputes at home, where Suzette, now twelve, kept sparring with him at every turn, challenging his authority and mocking his faith.

What he did—perhaps by way of revenge—was to lavish on me all the tenderness and affection he was thought incapable of giving. And I, left in the care of the Captain, basked and reveled in his love, found his quiet smile more comforting than my mother's elegant phrases, and experienced none of the harshness and imperious manner that had so alienated my older siblings, in particular, my sister.

I was barely six weeks old when the British and the French and the Israelis pounded Egypt with their rockets. César was placed in charge of securing both the house and me. He made sure all the wooden shutters were bolted shut, and then carried the crib to the middle of the cool, dark bedroom at the back. Loud, jarring air-raid sirens were frequently sounded, a signal to turn out all the lights and lie low. The explosions were only a few kilometers away, yet we could still occasionally see planes in the sky, and César, from his perch by the window, watched the clouds of thick black smoke in the distance. The sirens were sounded again once a raid was over, and my family huddled for safety in the back room of Malaka Nazli.

I could not stop crying.

The turmoil in Egypt cast an enormous shadow over my arrival. A few months before my birth, Nasser defied the world by nationalizing the Suez Canal, an act that enraged the British and the French. Nasser's rhetoric against Israel grew ever more belligerent, and he sent fedayeen guerrillas to launch attacks inside the country. The Israelis, convinced he was planning to invade, decided to strike preemptively. They found a willing ally in Britain's prime minister, Sir Anthony Eden, who had compared Nasser to Mussolini and even Hitler, and was passionate about the need to overthrow him. In a secret meeting, Sir Anthony, the French prime minister, Guy Mollet, and Israel's David Ben-Gurion agreed to send their armies into Egypt, take back the canal, and topple Nasser.

The world seemed to be coming to an end. Within a week of the assault on Egypt, Soviet troops moved into Hungary, using a massive show of tanks and soldiers to crush a revolt against Communist rule. Despite their own brutal invasion of another country, the Soviets deftly drew attention to the Western show of force in the Suez, loudly condemned it, and threatened to attack London and Paris. They were prepared to use "modern weapons of destruction" if there wasn't an immediate pullback.

President Eisenhower abandoned his traditional allies and joined the Soviets in demanding an immediate cease-fire in the Middle East.

Prime Minister Eden, shocked and profoundly humiliated, withdrew his troops within days, as did the French. But it was Eden's capitulation in Suez that had the most profound historical significance, for it meant both the end of the British prime minister and the end of the British Empire.

The Suez war, which ended on November 6, led to convulsive changes inside Egypt. Anyone holding British or French passports was given as little as forty-eight or seventy-two hours to leave the country. Families who had lived in Egypt for generations, whose children were born there and knew no other way of life, were escorted to the airport and, as squads of rifle-toting soldiers watched, put on planes bound for Europe.

Nasser's speeches brimmed with venom. He vowed to rid Egypt of all "foreigners," to eliminate the Jewish state, and stamp out the last vestiges of colonialism and the monarchy. People lost their jobs and livelihood overnight when the regime sequestered a business, placing Nasser's officers and loyalists in charge, and insisting that most employees be Egyptian nationals. Entire industries were nationalized as Nasser moved closer and closer to the Soviets.

At the movies, during intermission, theaters showed reels of Nasser and Om Kalsoum toasting the defeat of the European and Israeli marauders. The singer was one relic of the Farouk monarchy who was thriving. Because of her overwhelming popularity with the Egyptian people, Nasser courted her and even professed to be among her biggest fans. These days, the woman my father had adored, and who was said to have been his mistress, delivered fiery anti-Israel tirades and positioned herself as a true daughter of the revolution.

The Jews of Egypt followed the swirl of events with a sense of profound foreboding. The charmed life they had known under Farouk and his late father, King Fouad—the sense of being one of the most cosseted and most privileged Jewish communities in the world—was coming to an end.

The repercussions were felt in every household, including mine. My siblings all attended French schools that found themselves suddenly bereft of much of their staff. At the Lycée Français de Bab-el-Louk, where my sister was a student, Suzette had taken gym classes where she learned to dance the quadrille, a graceful eighteenth-century French square dance. In the aftermath of Suez, the school instituted special military training courses for its girls to teach them to fight against the Western—and Jewish—invaders. My sister and her school-mates were handed cumbersome old rifles, some dating back to World War I, and taught how to aim and shoot. At the Collège Français, my eleven-year-old brother César was mastering the same military skills, practicing on a rifle he could barely lift.

There was also a strong new emphasis on learning Arabic. In prior years, Egyptian girls, if they were members of the privileged class, disdained their native language and showed off their wealth and refinement by conversing only in French or English or Italian—any language but their own. But now at the lycée, Suzette found that she had a heavy Arabic courseload, in keeping with the ethos of the revolution.

My sister embraced the fitness program and learned to sing the new Egyptian national anthem with gusto. Contrarian to her innermost core, she sided with Nasser and told anyone who would listen, including my startled family, that the British and the French were wrong to have invaded and that the canal belonged to the Egyptians.

But even she wanted to leave: all the young people realized that they had no future in Egypt.

The grand synagogue on Adly Street became a hub of frenetic activity, the scene every day of hurried weddings. As families prepared to flee to any country that would have them, as they plotted their escape literally to the ends of the earth—Australia, Venezuela, Canada, South Africa, Brazil—young lovers chose to tie the knot lest they be separated forever. Engagements that would have lasted months were now barely a couple of days, while weddings that usually took a whole evening were performed in an hour.

Rabbi Chaim Nahum Effendi, Egypt's venerable chief rabbi, found himself officiating at multiple ceremonies in a single day. These were assembly-line weddings, with little of the pomp and ceremony that had marked a traditional affair at the Gates of Heaven.

Young brides hugged their parents good-bye and took the first boat out.

There wasn't even time to cry—there was only a feeling that one had to get out at any cost.

Egypt had witnessed this kind of hysteria before, in 1948, during the first Arab-Israeli war, and then in January 1952, after the Day of the Four Hundred Fires. Somehow, the sense of hopelessness and finality seemed more intense in the aftermath of Suez.

It was no longer a question of whether to leave Egypt, but when.

MY FATHER URGED EACH one of his brothers and sisters to go immediately to Israel, the one country that would take them, no questions asked.

But at home on Malaka Nazli, my father made no move to travel anywhere. "C'est à cause de Loulou," he'd say if anyone asked. We would leave too, but for concerns about my welfare, he explained. After what happened to Baby Alexandra, he wasn't prepared to put his delicate infant daughter in a perilous situation, taking her on a long sea voyage she might not be able to survive, not to mention resettlement in a primitive country like Israel where it was whispered that new arrivals were put up in tents or metal barracks that were as hot as ovens when the sun shone. Still, he was committed to abandoning Egypt "quand Loulou est un peu plus grande"—when I was a bit older.

It would be only a matter of months, he vowed.

The patriarch had spoken. Members of my extended family prepared to embark on what was, in effect, a second Exodus. One by one, all my aunts and uncles—Marie and her six children, Oncle Raphael, his two beautiful daughters and lone son, Oncle Shalom, Tante Rebekah, her husband and one of her sons, were set to leave for the Promised Land.

"On va se revoir bientôt—Inshallah," my dad vowed.

Then it was Alexandra's turn to leave.

Alexandra's support systems were vanishing one after the other, like the cigarettes she chain-smoked. The cousins and second cousins who had kept her out of abject poverty—or allowed her to survive in spite of it—were departing in droves. The *communauté juive*, with its chaotic but fairly generous charitable network, on whom she could count for the occasional handout or a meal, was also being dismantled. All the major donors, as well as the minor donors, were disappearing. The *Hôpital Israélite* was shut down in 1956, and it was all the community could do to ensure that the Jewish home for the aged in Heliopolis remained open, and to donate money to the Yellow Palace, the state insane asylum, so that Jewish patients would continue to receive care.

Alexandra's only real option seemed to be her son Félix in Israel, who sent word she could join him. But no one had any illusions about Oncle Félix, the charming huckster and ne'er-do-well who had stolen my mother's engagement ring and pawned the precious stones on the eve of her wedding, who had shown himself unable to keep a promise or commitment his entire life.

No one had any illusions, that is, except Alexandra. She would have loved to move in with us, of course, but she was so timid and unassuming she wouldn't have proposed it on her own, and my father wouldn't have stood for it, and my mother was too weak to insist on it, and my sister was too young and too frazzled, her anger too diffuse, to demand it, the way that she alone seemed capable of standing up to my father.

Alexandra gave up her small room and her favorite coffeepot and moved in with us while her papers were finalized.

Since all the beds were occupied, she was offered a trunk in a corner of the living room, cushioned with blankets and pillows. And there Alexandra slept in her final days with us, a small frightened woman on the eve of a journey she had no wish to undertake, lying on top of a large old-fashioned black trunk made up to look like a bed.

A few weeks later, she left for Alexandria, where she was met by relatives—the wealthy branch of the once-grand and noble *famille Dana*. They, too, were hurriedly planning their escapes, but in one final act of charity—for they were renowned as great benefactors of the city—they prepared a small suitcase for my grandmother, and stuffed it with whatever they could gather: bits of clothing and lingerie, a couple of books she would never read because of her cataracts, several packs of cigarettes, biscuits, hard-boiled eggs and tins of cookies lest she go hungry on the boat.

My family went to see Alexandra as she prepared to depart. Suzette was shocked—my grandmother's shiny black hair had turned white. It had changed color in the time it had taken for her to travel from Cairo to the city she had abandoned in her youth. Alexandra was assured we'd be joining her any day.

Perhaps she believed it—perhaps we all did. It seemed unthinkable that we would never see each other again, that families would go off in a thousand different directions, like the ships setting off from the port of Alexandria.

The dock was impossibly crowded, as if all the remaining Jews of Egypt had chosen that moment to go. The thin old woman with the white hair tied neatly back in a bun was allowed on board without a fuss. Nobody offered to help her, or to carry her luggage. Chivalry, once an essential ingredient of life in Egypt, had also disappeared, as if it too were a relic of colonialism that had to be eradicated.

Alexandra of Alexandria was gone, an old woman, nearly blind, trying to hold on to her heavy valise on a crowded boat.

My sister, who would grow up to marry a man named Alex and would name her only child Alexander, was inconsolable. She left us and wandered off to a local cinema, where she bought a ticket for a couple of piasters. The theater was playing *Love Me Tender*, which starred the new American heartthrob Elvis Presley. Suzette cried from the first scene to the last. Seated alone in that cool darkened theater, as Alexandra had always done, my sister cried and cried watching Elvis and listened to him sing the lyrics to the song that she would forever associate with that awful day in the summer of 1957.

"Love me tender, love me dear, tell me you are mine. I'll be yours through all the years, till the end of time," Elvis sang, as my sister thought longingly of our grandmother, and wondered how she was faring on her choppy and hopelessly solitary journey.

CHAPTER 7

Alexandra in the Promised Land

Once in Israel, my relatives' worst fears were realized. Some were placed in rugged settlements in the middle of the desert or in remote agricultural areas where home was a tent, an armylike barrack, or a flimsy structure made of aluminum. Tante Marie, appalled by the squalor of their new life, wondered why on earth she had listened to my father. Even in their diminished circumstances and with the convulsive political situation, Egypt had been better than this.

And where was Leon? Why had he sent them to this wilderness?—which is what she considered Kibbutz Givat Brenner.

My dad's letters to them were both reassuring and noncommittal. My cousins were living in a primitive prefab dwelling made of aluminum siding; the rooms were unbearably hot. Their diet was spartan, the opulent dinners of chicken and meat they had enjoyed in Cairo a wistful memory.

The only solace came from the person my father most hated, my lost uncle, Salomon the priest. Shortly after arriving, there was Père Jean-Marie, come to comfort his family from his monastery in Jerusalem. He was like a vision, striding through the fields in his long black priestly garb, and the residents of Kibbutz Givat Brenner could only watch with amazement as the stocky, well-built man with a full beard, a large cross around his neck, and one dangling at his waist, made his way through their commune.

The kibbutzniks were even more astonished when the object of his visit turned out to be the desperately poor family of refugees from Egypt, in particular, a small, pudgy woman who couldn't speak a word of Hebrew and who hadn't stopped crying from the moment she'd reached Givat Brenner.

For Tante Marie, who felt as if all hope had gone from her life since leaving Egypt, being reunited with her older brother was a miraculous event. She saw him as a saintly figure come to show them kindness and compassion amid the bewildering harshness of their new home.

Oncle Raphael was in even worse straits. His health began to fail. Without my father, his trading partner in sardines, jams, and olive oil, to comfort him, still heartbroken by the events in Egypt, Oncle Raphael suffered a heart attack and died. He had been in Israel only six months.

Shortly after arriving in Israel, Tante Rebekah also became ill. She was diagnosed with lung cancer, though she had never smoked a cigarette in her life. Her son David, an enlisted soldier in the Israeli Army, came home to help care for her. Her husband knelt by her bed as she lay dying.

And then there was Alexandra.

My grandmother had landed in the middle of the orange groves of Ganeh Tikvah, an agricultural settlement. It could have been the end of the earth, this desolate patch of nothingness (but lots of oranges) miles from the nearest town. But it was where her son, Oncle Félix, lived, working when he could as a journalist, and it was supposed to be her new home.

In fact, she felt more like an alien from a distant planet. She didn't know a soul, couldn't speak the language, and seemed incapable of finding her way around, though even if she had, it wouldn't have mattered because there was nothing to find.

Home was a narrow bed in the wooden prefab dwelling where my uncle and his wife, Aimee, lived. They were at war, trading barb after barb, accusation after accusation, with my grandmother as a witness.

Alexandra did what she'd always done to escape a painful situation— she went for walks. She left her son's wooden house and wandered up and down the narrow gravel walkways that were the closest approximation to streets in this godforsaken village, where all one could see for miles and miles were orange groves.

She walked so much, and her manner was so distracted—mumbling to herself, holding on to a cigarette butt—that she attracted the attention of this immigrant community, made up mostly of refugees from distant corners of the Levant—Tripoli, Tunis, Algiers, Casablanca.

They were sure that she was mad, this hunchbacked old woman who was always muttering to herself as she paced up and down on the same narrow road, and who would occasionally stop and address them in a language they couldn't understand. She seemed to be pleading for help but didn't know how in Hebrew or Arabic.

Alexandra became a figure of pity, or worse, derision. Children made fun of her and laughed in her face, and their parents weren't much better. In the hardened society that was Israel of the 1950s, Alexandra's fellow Jews showed her far less kindness, less of that wondrous quality the Egyptians call *rahma*—mercy, compassion—than she had encountered in Cairo at the hands of Arabs who'd see her wandering around Malaka Nazli.

No one had laughed at Alexandra in Egypt. On the contrary, strangers, moved by her air of distress and helplessness, would often stop and ask her if she needed help. That was the difference between a society that was primarily Middle Eastern, despite its long colonial history, and one that was primarily Western, despite its geography, and longed to be even more so.

My grandmother wasn't crazy—she was simply lonely and feeling horribly unwanted and unloved. When were Edith and the children coming? she wondered. When would Leon make good on his promise and move the family here, so they would all be together again?

Though she lived with family, she was, for all intents and purposes, alone. She didn't have a dime, and her son didn't either. Oncle Félix hadn't changed a bit between Cairo and Ganeh Tikvah. Though good-natured, he was still undependable. Almost from the moment my grandmother arrived, he'd announce he was going away, and off he'd vanish—to the other end of Africa, or merely to the other end of town.

Back in Egypt, Alexandra's future had seemed settled, if not exactly ideal. Oncle Félix was said to be waiting for her in Israel in a house set amid farms and bucolic fields. But despite its rich soil and delicious outsize fruit, Ganeh Tikvah was a wasteland as far as Alexandra was concerned, a village with no shops, no theaters, no life—only one small grocery store. It wasn't a kibbutz, so there weren't even communal activities that could have lessened her sense of solitude. She couldn't speak the language, she didn't know the area, and she didn't have any pocket money.

And she hated oranges.

How different from Cairo, when even at her lowest ebb, Alexandra could step out of her rented room and find companionship and cheer wandering amid the maze of boutiques and stalls and cafés and restaurants and cinema houses—above all the cinema houses. Not simply the Rialto, her own small haven where even the ticket clerks were her friends, but also the sumptuous *cinémas en plein air*, outdoor theaters like the Rex and the St. James where it was possible to sit on a wicker chair and feel the cool breeze from the Nile while enjoying a double feature.

She had been enthralled with the new generation of stars of the 1950s—Van Johnson, Deborah Kerr, Rock Hudson, Grace Kelly, and Elizabeth Taylor, always Elizabeth Taylor. How she endeared herself to Suzette by comparing her to Elizabeth Taylor. How she'd loved going to Malaka Nazli and telling Edith's children the plotlines of the latest films she'd seen—embellishing them, of course, but only a little bit.

The air of Ganeh Tikvah was heavy with the fragrance of orange flowers. Once upon a time, she had loved to sing, "Senti, senti questi fiori d'arancio"—Inhale, inhale the orange flowers; but now she felt she was suffocating, that she couldn't escape the trees with their immense fruit and the scent that was everywhere.

How she missed her old room and her *tanaka*, the little coffeepot that was her closest companion.

FINALLY, MY GRANDMOTHER MADE a friend of sorts. Newly married, eighteen-year-old Josette also was feeling lost and homesick. She missed Cairo and its comforts, and her parents who had stayed behind. She thought of Alexandra as an emissary from civilization. At last, someone to share a *café Turque* and enjoy a pleasant chat in French with—no Hebrew or Arabic, please.

Josette lived with her husband in a small wooden cabin in Ganeh Tikvah that Alexandra frequently passed on her aimless wanderings. She was actually related to my grandmother and Oncle Félix through marriage. Her husband's grandfather was Isaac, the man who had wed, then abandoned, Alexandra, and his mother was Tante Rosée, my mom's half sister. It made sense that, once in Israel, Josette and her husband had reached out to Oncle Félix.

In spite of her abject poverty, Josette could tell that my grandmother had that indefinable quality that could only be called class. She had an aristocratic bearing and her French was so exquisite! She was also a natural raconteur, and much as she had once beguiled my siblings, she captivated the young woman. With her exquisite French, Alexandra brought to mind some of Josette's former schoolmates at the Lycée Français de Bab-el-Louk, the fabulously rich girls who arrived each morning in chauffeured Cadillacs from their mansions in Garden City and Zamalek and whose servants brought them lunch. What Josette couldn't have known was that once upon a time, Alexandra had been one of those girls.

She found the old woman endearing, the contradictions in her both haunting and fascinating.

On the one hand, she was so helpless, the most helpless person she had ever known. In Israel, even more than in Egypt, my grandmother didn't have a clue how to cope with the ordinary exigencies of life.

Alexandra was like a flower, Josette decided. But not any flower—the rare, delicate variety that grows on the side of a mountain, the edelweiss, white and lovely and excruciatingly fragile.

There was another quality she couldn't pinpoint, which was wrapped up in the secret sorrow of the woman. It was an aura of mystery, a sense that she was searching, searching as she walked up and down those narrow pathways shaded by the orange trees, but for what, for what? "She is looking for her son," the young woman concluded. She hated Félix and blamed him for Alexandra's unhappiness.

When Josette's own parents finally moved to Israel, they settled in an adjoining wooden cabin and instantly embraced my grandmother. Alexandra would wander over to their house during one of her endless walks to nowhere, the way that she had once stopped to see us at Malaka Nazli. Josette's parents were amiable people. They learned to recognize the shy knock on the door—four taps, and there was Alexandra, cigarette in hand, a slight smile on her anxious face.

She was so painfully thin by then. She was probably starving. There were ripe, delicious oranges literally at her feet. Pieces of fruit littered the grounds of Ganeh Tikvah, and on any one of her walks, Alexandra could simply have bent down and scooped them up. But she never did. She survived, if barely, on a diet of cigarettes, Turkish coffee, and the occasional hard-boiled egg that she ate every couple of days. Though she was delightful company, she seemed terribly distracted, unable to sit still for more than a few minutes, in a hurry to get to her next destination.

"Tu ne veux rien manger, chérie?" Josette or her mother would ask, Won't you have a bite?

Alexandra would shake her head no, and politely excuse herself. She left them to resume her walk, which was both frantic and aimless.

Josette would watch her from the window—a little girl lost in the guise of an old woman, oblivious to her surroundings.

La guigne—bad luck, the evil eye. They had haunted my grandmother her entire life and had followed her here in God's country, the one place on earth where curses could be broken and destinies reinvented.

"Please, God, don't let me end up like that," Josette prayed.

My grandmother was in fact intensely aware of her surroundings. She was also—as Josette had guessed—on the lookout for her son.

But not *that* son, not my uncle Félix. It was her other son, the blue-eyed child of the souk, that Alexandra hoped to find amid the orange groves. Surely her life couldn't end like this, in a desolate stretch of nothingness, penniless and unwanted and alone, without her even having solved the essential mystery that had haunted her these many years. In my grandmother's anguished solitude, being reunited with him had become her obsession once again.

He was all grown up now, but she was sure as only a mother can be that he had left Egypt too, and was somewhere close at hand.

She was persuaded that he too had found his way to the Promised Land.

Ganeh Tikvah, after all, meant the Garden of Hope, and Alexandra kept walking on the dark gravel roads, determined to remain hopeful.

CHAPTER 8

The Arabic Lesson

I am always drawn to photographs that seem to foreshadow events— happy scenes that contain telltale signs of a tragedy still to come. Everyone is smiling, but somewhere on the edge of the frame is a dark smudge, a shadow, so that despite the joyful faces, it is clear that a terrible event is about to unfold.

No matter how often I've stared at the last photograph of my father in white sharkskin, I have found no such hints. There he is in the courtyard of Temple Hanan. The occasion is a bris, a circumcision, and several guests have gathered in a large semicircle in front of the graceful Italianate windows. They beam in their soft, flowing silk dresses and tailored suits. At the center, my cousin Edouard is by his wife and mother-in-law, who cradles the newborn in her arms. My dad, though off somewhat to the side, still stands out because of his tall, princely bearing, his confident, easy smile, the fact that he towers over everyone around him, and mostly, his breathtaking elegance.

All the men are wearing dark, conservative suits.

Leon alone is in white.

It is a blissful scene. I have always thought of it as the last happy

The last happy picture of the Jews of Egypt. Temple Hanan, Cairo, 1958.

picture of the Jews of Egypt. Nothing portends of the events to come. There are no hints that within weeks or months it will all be over. The baby's parents—Edouard, my father's cousin, and his wife— will take their young son and depart almost immediately for America. The women in their finery will be dispersed to a dozen foreign lands. Temple Hanan will be abandoned, its courtyard empty and forlorn. And my father will never dress, or stand, or smile quite this way again.

We didn't go to Alexandria in the summer of 1958, a break from our tradition of renting a house or apartment by the sea. Life had lost much of its luster after the mass exodus of our relatives. Our house felt forlorn, devoid of its usual visitors. My father fielded letters from Israel, asking when we would be joining the rest of the family.

But after the panic that followed the Suez crisis, there was the semblance of stability. Life seemed to resume its languorous Levantine pace—at least on the surface. Though Father kept reassuring our relatives we were planning to join them, there didn't appear to be quite the same urgency.

Even so, a lingering effect of the 1956 war was fear, a nagging sense that the Nasser regime was spying on us, that danger was around the corner. The maid, the porter, the street vendor—anyone could be spying for Nasser's henchmen. My mother would sometimes motion to my siblings to keep silent: "Les murs ont des oreilles," she would say, The walls have ears, and point to the maid setting the table.

Many people we knew had lost their businesses to the government, but Leon had almost nothing the government could sequester or confiscate. And so, in a strange way, he was able to continue as he always did—working resolutely alone. He outwitted and outmaneuvered the regime so that it couldn't seize what was ours. Nasser himself couldn't have penetrated the layers of secrecy shrouding my father's myriad business interests.

My father's routine remained unchanged. He still woke up every day at sunrise to attend services. Afterward, he came home for a light breakfast, then left again to go downtown for meetings with clients. He'd stop to enjoy a glass of cold beer, wander over to the bourse and more business meetings.

He loved to walk across Cairo, and he was so vigorous that even now, nearing sixty, he didn't feel a need to slow down. Besides, he was energized by all his responsibilities at home, the fact that he had four small mouths to feed, including me, the new baby.

I was, from the start, his personal charge. When he was home, Edith left me in his care and together we went to play in the park across the street, on the campus of Sacré Coeur, or we'd simply stay in his room facing Malaka Nazli; as he worked or leafed through the morning papers, I sipped on milk fresh from the cow that still came to the back of our house each morning, exactly as it had in my grandmother Zarifa's day.

One morning, cousin Edouard walked by and spotted me in my father's arms, playing on the windowsill.

"Bonjour, Captain, ça va?" he called out; All's well?

"Dieu est grand." My dad smiled.

Edouard was trying to mend fences with my father. Shortly after his son's bris, word spread he was planning to live in America, a move he'd tried to keep secret until the last minute. Though most of his friends approved, my dad was furious. Now he had to confront the Captain's full wrath. Edouard's father was sick, a patient at the old-age home in Heliopolis. How could he abandon him?

My father urged Edouard to stay calm and reconsider: there was no immediate need to leave Egypt now. But my cousin seemed panic-stricken. In his mind, he had no choice but to abandon his home, his job, and his life in Cairo. His wife was adamant; even if life was safer than after Suez, Egypt still held no lasting prospects for Jews.

Another morning, my dad summoned a photographer to come take a picture of the two of us outside Malaka Nazli. I toddled alongside him across the wide boulevard, over to the Convent of the Sacred Heart, whose grand, imposing building made such a perfect backdrop. Neighbors and passersby watched as the photographer set up his tripod and old-fashioned camera in the middle of the street.

My dad spotted a graceful old automobile, a Sheffield, parked in front of Sacred Heart. He picked me up and positioned me on top of the hood. He leaned over, pointing to the camera, hoping he could coax me to smile. He couldn't. I did obey him in one sense, by staring, wide-eyed and unblinking, directly toward the photographer, who promptly poked his head under the black curtain and cried out, "Par-fait."

In the resulting shot, Dad holds me in that tender, protective way that is uniquely his. Though I seem a bit befuddled, he smiles broadly enough for the two of us. He is dressed somewhat informally, in a shirt and tie, but no jacket; I wear a prim cotton dress and real leather shoes, not baby boots, and my hair is styled in a perfect pageboy. The camera captures it all—the joy he has in holding me, the sense of absolute safety I feel nestled in that favorite spot I've staked out in the nook of his shoulder, *au creux de son épaule.*

It is a universe of two, created here on the streets of 1950s Cairo, and it will always be like this, my father and I taking on a vast and

Leon and Loulou.

difficult world together, with a swagger and a smile and an expensive car we don't even own. Nothing in the photograph suggests it will ever be any different, that the idyll will ever end.

It did, and only a couple of weeks later.

My father woke up at four in the morning as was his wont, determined to make the first service at Temple Hanan. He put on his lightest, whitest clothes because it was so brutally warm out, one of the hottest summers he could remember.

He began the familiar ten-minute sprint to synagogue, crossing the alleyway by the side of our house, then turning toward a main street which led to Midan Sakakini, the wide traffic circle anchored by the mansion of the fabled Sakakini family of pashas and beys, then back to the narrow dirt roads he knew so intimately that he could have walked them blindfolded. He had been taking the same familiar path every morning for more than twenty years, since he'd moved with Zarifa to the house on Malaka Nazli.

PERHAPS IT WAS A manhole. Or a crack in the poorly paved road. A rock. A slippery patch of cement.

He would always ask himself, and he would never be sure.

Suddenly he found himself falling, falling.

He flew through the air and hit the pavement with such force he was sure that all his bones were broken and that he would die. That was the problem with being tall and well built—he collapsed with unearthly speed and an enormous thud. He couldn't move, he could only lie there moaning, while passersby swarmed around him.

In old Cairo, no one kept walking if you were hurt.

He was unconscious when the ambulance arrived. The crew rushed him to the Demerdash, one of the closest area hospitals. The Demerdash was a public institution that took care of Cairo's poorest citizens, the indigent, the fellahin. My father would always say that it was God's will that he was brought there.

At home, the call came later that morning. My mother screamed. She ran to the *ba-wab* to summon a taxi. She seemed so frail as the porter helped her into the cab, and she was trembling. In the background, my brother and sister were crying.

At the hospital my father fleetingly regained consciousness. He thought back to his life, to his all-consuming love for God, the fact that he had devoted himself to prayer and charity and acts of goodwill, and wondered why he was being punished in this way, why he of all people had to endure this excruciating pain.

When he cried, he asked for one person: Zarifa, my grandmother, dead for more than a dozen years.

Malaka Nazli alone absorbed the news coolly, impassively. She had been a house of tears before, and she was prepared to accept her fate again.

DR. AHMED KHATAB RUSHED to my father's bedside. The surgeon shook his head in dismay as he delivered the news: my dad had smashed his leg and broken his hipbone—it was completely shattered because of the brutal impact of the fall. He had to operate immediately to try to repair the hip, though he wasn't overly optimistic.

The Demerdash was a little shack of a hospital where the poorest of the poor went for care. From the start my family agonized about whether my dad should stay there. Across the road was Dar-eh-Shefah, a sparkling private medical center that was said to house some of the finest doctors in all of Egypt. It had modern operating rooms and top experts in orthopedic surgery, and it made eminent sense for my dad to be transferred there at once.

He wouldn't hear of it.

If God had brought him to the Demerdash, he told my mother, that is where God wanted him to stay.

He was simply adamant, and there was nothing anyone could do. He certainly had the means to go to the finer hospital across the road, or anywhere else in Cairo. But there was no persuading him, though my mother tried and tried. Nor was she assertive enough to simply insist on it.

Dr. Khatab performed the operation, using the most modern techniques he could muster. To repair Dad's crushed hip, he inserted a hard metal pin known as a Smith-Petersen, which had been developed at Harvard by an eminent surgeon of the same name. My dad was under anesthesia, but for years, even decades, thereafter, he could recall with a shudder the pounding pounding pounding pounding of the hammer used to drive the Smith-Petersen nail into him.

Then there was the pain—the pain that was constant and unremitting.

Each day, my mom would take a tram or taxi to the hospital. She didn't know quite what to do with me, so she often left me with my sister, since Suzette didn't much care for going to the Demerdash, and at fourteen, her anger toward my dad was undiminished. When she and my mom spoke about the accident, it was to bemoan the fact Dad had insisted on getting up at these unearthly morning hours to go to synagogue, when other devout men didn't feel a need to do as much. "C'est simplement du fanatisme," my sister said, and my mother sadly agreed. It was Dad's excessive, fanatic devotion to religion that had landed us all in this mess.

Occasionally, Mom took me with her, so that I got my first taste of hospitals when I was not yet two.

My father, who had always been so silent and stoic, now seemed agitated and perpetually distraught. He'd complain about relatively small matters, like the food, which forced us to come rushing over to the Demerdash with home-cooked fare in the little aluminum traveling caddy with pots containing his favorite dishes—a serving of stewed lamb, or stuffed grape leaves, white rice, and roasted potatoes.

He'd eat and momentarily calm down, but then fly off into a rage all over again, complaining about the nurses, the squalor of the hospital, the fact that the doctor didn't come as often as he wished, though he was fond of Dr. Khatab, who seemed devoted and stayed far longer than necessary by my dad's bedside trying to restore in him some sense of hope.

My father, who had defined himself all his life as a boulevardier, was despondent at the thought he would never walk again.

César was constantly at the hospital. He had taken up a new hobby, drawing, and to keep himself busy while visiting, he brought along his sketchpad and a set of charcoal crayons. One afternoon, as Dr. Khatab came to my dad's bedside, my brother began to draw the doctor. It was such a realistic portrait that Dr. Khatab offered to buy it then and there.

César sketched the nurses, the Arabic peasants floating down the hallways in their long, white galabias, a couple of the patients, and Dad. My oldest brother always had a quiet, soothing manner, but nothing he said seemed to calm my father, or reduce his existential anguish.

He knew, as we all did, that it wasn't only his leg that had been shattered that morning on the way to Temple Hanan, but his entire life. He had never been able to sit still, except, oddly, at temple, where he had such staying power he could remain seated in his favorite chair by the Holy Ark and pray longer and more fervently and with more focus than all the other men, even the rabbi.

At night, he was never tired enough or satiated enough to want to stay home. He'd head by taxi to his favorite nightspots—L'Auberge des Pyramides, the Mokhatam, the restaurant that doubled as a casino, or else one of the outdoor cafés where it was possible to have dinner and then go dancing, because he enjoyed dancing every bit as much now as when he was younger.

And now, it was over—a way of life he had loved so much that he had risked his marriage to sustain it.

He would never dance again, he was certain of that. He would never go into a nightclub and have a beautiful woman float into his arms and while the night away with him in a carefree rhumba or cha-cha-cha, an elegant waltz, or his favorite, an intense and passionate and carefully executed tango.

IT WAS LATE FALL when my father came home. He promptly settled in his white brass bed. Every day, a nurse would come over to try to exercise him, and Dr. Khatab dropped by almost as often to examine him and check on how he was healing.

He wasn't. There was so much pressure on the pin that it broke, and the surgeon insisted he needed another operation to remove the pieces. My dad, haunted by the memory of the hammer pounding the nail inside his body, shook his head emphatically no, no, no, no.

"Dieu est grand," he exclaimed, when the doctor asked him how he expected to get well without the needed surgery.

The Muslim surgeon nodded thoughtfully, respectfully—and resolved to try again. Failing to remove the broken nail could only lead to disaster.

Meanwhile, he prescribed a rigorous home regimen to stop the affected leg from shrinking. The idea was to keep it stretched, using a primitive homemade system of weights and metals, including the cumbersome heavy irons the maids used to press our sheets and linens, as a means to keep pulling on the leg.

The pressure meant that my dad was in a perpetual state of discomfort. In truth, everyone on Malaka Nazli was on edge. My mother was suddenly in the position of having to support and care for all six of us, and she had no idea how to go about it. My father had always been as secretive about his finances as about his work, and she knew nothing of bank accounts, stock certificates, ownership stakes, cash holdings. He had always paid all the bills, including the maids' salaries and my siblings' private school tuition, while giving my mom a small allowance for her personal needs.

In desperation, my mother confronted him. She had no money left, and could afford nothing, not even upkeep of the house.

"Je ne peux même pas payer le loyer," she said; I can't pay the rent. It was a bit of an exaggeration, since our rent was pitifully low, only a few Egyptian pounds, about a dollar or two, and it hadn't changed much since my dad had moved into the house with Zarifa and my cousin Salomone in 1938.

My father finally relented, and directed her to the *sandara,* the small crawl space on top of the kitchen. Every Egyptian house had one, and typically it was used for storage—of canned goods, as well as odd pieces of furniture, knickknacks, books, even old cooking pots and ladles left over from my grandmother's day and that Mom had put away after Zarifa died. Buried deep inside the *sandara,* Dad said, was an old briefcase where she would find cash as well as stock certificates she could go and redeem.

My mother climbed a small ladder, retrieved the ancient leather attaché, and was amazed to find the key to my dad's finances—stock certificates, bank accounts, as well as Egyptian pound notes in large denominations. She had been married to my father nearly fifteen years and hadn't even known of the bag's existence. Without checking back with him, she took what she needed from the bundles of Egyptian pound bills, pulled out the necessary stock certificates that she planned to cash, placed the briefcase back in its hiding place, and returned to my father's side.

We had enough to live on till he was well again.

MY MOTHER WAS NOW in charge. She made all the decisions, dispensed all the money, paid all the bills. But instead of empowering her, the surfeit of responsibility seemed to make her more and more anxious and abrasive. It was as if, stripped of her traditional role as subservient wife, she couldn't handle the pressure and became much like that tough, domineering man she had bristled at all these years.

Oddly, it was twelve-year-old César, the model son, the favorite, who bore the brunt of her rage. He was still too young to assume the role that would come so naturally to him in later years, when he would be the de facto head of the household. Now, he simply retreated to his drawings. He drew anyone who floated into Malaka Nazli—the maid, the porter, friends of my father from the different synagogues he attended, my dad in his sickbed, Dr. Khatab paying us a house call. Seated by the window of my father's room, he also sketched landscapes from memory—iconic images of Egypt like the Great Pyramid at Ghiza, or the Sphinx, or feluccas drifting down the Nile.

His portfolio grew so thick that in the months of my dad's accident and excruciating convalescence, he had several dozen finished portraits etched in black and white.

It was a way to forget, to escape the cries that now routinely filled the house.

One day, as my brother was sitting in the large back bedroom, quietly drawing, my mom came in. She was cross, as she often was, these days. César decided it was prudent to ignore her, but she would not be ignored. She asked him why his bed was so messy. Why was he drawing when he still had homework to do? Didn't he realize how much she was paying to keep him enrolled at the Collège Français?

César still didn't answer her, and kept his head down, focused on his drawing.

It all happened so fast, like my father's fall.

My mother suddenly grabbed the portfolio and began tearing it up. It didn't matter that he kept crying to her to stop. She wouldn't listen. One by one, she destroyed every single drawing that he had worked on so painstakingly—the charcoal portrait of Dr. Khatab, the bucolic scenes of the Nile and the Pyramids, the sketches of the maid and the washwoman and the porter, and the cat, and of Dad looking longingly from his bed toward Malaka Nazli.

LITTLE BY LITTLE MY father began to walk again. He would take small, unsteady steps around his room, then shuffle through the rest of the house, but it was a year or more before he went outside, and then only very tentatively and for the shortest of distances.

His first destination when he could move around by himself was the synagogue. That had not changed.

What had changed was how much we now worried.

Each time he went out, César would position himself on the balcony, waiting for our father to return. He watched him walk down the alley till he was out of sight, then remained in place to see him safely return. My brother had a nagging fear Dad wouldn't come back, that he would fall down again. He seemed so precarious and fragile with his wooden cane. The confident bearing was gone. He seemed less tall, perhaps because he now stooped as he leaned on the cane. Overnight, it was as if he had aged by years. He didn't go out very much, and only by day; when he did, he came back early and went to lie down in bed. There were no more casinos, or nightclubs, or friendly games of poker, or women.

Somehow, having my father home all the time didn't seem to make Mom any happier than when he'd rarely been around. Still, even frail and infirm, he made the perfect babysitter. My mother realized she could leave me with him and go about her day and run errands and see friends at Groppi's, and shop and consult *la couturière* about a new dress, without worrying about either of us.

Watching over me was also a distraction for my father. He had always been partial to very young children, so that he didn't view taking care of me as a burden but rather as a pleasure. From the start, we were entirely compatible. It was as if even as a toddler, I could sense his anguish, and tried to be on my best behavior.

I was, by nature, a quiet, placid child. I had no interest in playing with other children; I was content merely to sit by his side day after day. When he was strong enough to move from the bed to a chair by the window, I followed him, sat on his lap or on a pillow on top of the windowsill, and Malaka Nazli became our personal theater, a source of constant diversion. If Dad didn't succumb to despair, it was surely because of the curative powers of the street, its life and energy.

There was no talk anymore of leaving Egypt. Now, of course, he couldn't go even if he'd wanted to, not with his bad leg. Yet nothing had really changed. Jews were still streaming out, driven less by a sense of panic than a sense of fatalism.

Life as they had known it was over. Egypt—Jewish Egypt—was finished and would never be again.

———

BEFORE I TURNED FIVE, Mom enrolled me in the Lycée Français de Bab-el-Louk. I lasted all of one day in kindergarten. My mother decided that the childish games and finger painting were a waste of time, that I was far too advanced for my age. During his convalescence, my father had taught me how to count to ten by using an object near and dear to his heart: a deck of cards. He would flash an image of two diamonds and I'd cry out, "Deux," or four pretty red hearts, and I would instantly say, "Quatre." Sometimes my sister came in to watch, because even she was amused by my father's unorthodox teaching methods.

My mother, citing my rudimentary knowledge of arithmetic, persuaded the lycée to have me skip kindergarten.

The lycée obligingly placed me in first grade, where a bilingual program of studies was already in place. Since the war, learning Arabic had become so important that even the youngest children were expected to master it. The problem was that I was hopeless at Arabic. I couldn't even manage to memorize the little Arabic folk songs that made up much of the curriculum.

Reports began filtering home from my teacher, who sent word that I wasn't grasping the most basic Arabic words, that my pronunciation was hopelessly flawed, and I couldn't deliver a simple greeting, or follow the rest of the class in singing.

My failure in first-grade Arabic threw the family for a loop.

Mom had done a splendid job teaching me French, using her skills from her days at the École Cattaoui. Though born in Cairo, I may as well have been one of those colonialists who were fleeing. My family dubbed me "Hawagaya," the Foreigner, because I spoke Arabic in the stilted, lispy, hesitant manner of a girl from abroad.

My Arabic teacher was honestly at a loss. She told my family they'd have to take responsibility, or she'd give me a failing grade.

It was my first of several identity crises. In Egypt, I was called a foreigner because of my inability to speak Arabic. In France, where we'd briefly sojourn, and where I was completely fluent in the language, I was a foreigner, because I was from Egypt. And in America, I was still a foreigner, because I came from Cairo and Paris.

My destiny seemed preordained: I was to be the perennial outsider, a *hawagaya,* no matter where I lived in the world.

Of everyone in my family, my father was the most perturbed. Arabic was his first language; the notion that a daughter of his couldn't master it was unthinkable.

He decided to take matters into his own hands. Every day after school, he told me to report to his room for private tutoring. He was usually sitting up in bed, praying or reading the newspapers. I'd climb up and sit next to him, my back propped up, like his, against several big feather pillows. He'd start each lesson by having me introduce myself.

"Esmi Loulou," I learned to say; My name is Loulou.

He would take my schoolbook in hand and begin the drill. I noticed that he held it as carefully and respectfully as his favorite red prayer book.

He'd point to certain words and have me read them out loud, then gently correct me when I was in error, and ask me to repeat after him.

He was the most mild-mannered of teachers, and the most focused. Unlike my mom, he never became angry or annoyed when I made a mistake. He seemed to have an infinite amount of patience. No matter how often I stumbled, he'd simply nod and have me go on. And I was always getting bonbons from the seemingly infinite stash he kept by his bedside.

My favorite times were when we put aside the book, and tried a more hands-on approach. Dad wanted me to learn the names of all the objects in his room, or even some that suddenly materialized, like Pouspous.

"Otah," he'd say, pointing to the cat, who had wandered in no doubt expecting a treat, not a tutorial.

"Otah," I'd repeat, lifting her up to the bed to join us. I wanted Pouspous to study Arabic, too; my father agreed that was a fine idea.

Only Pouspous seemed somewhat dubious, at first.

It was a cinch for me to learn the word for *cat,* and *pretty,* too. I learned to tell Pouspous that she was very pretty: "Otah helwah." Dad offered her a piece of cheese, one of her favorite snacks. She reached for it delicately with her paw, but before he let her have it, I had to show them both I'd mastered the word for *cheese.*

Pouspous looked at me anxiously, as I strained to remember.

"Gebnah," I finally cried out. My father smiled, and the cat grabbed the cheese and polished it off, then stood at attention, waiting for a few more tasty Arabic lessons.

If Dad was in a particularly good mood, he'd decide we should all feast on sardines. That was a freebie, because the Arabic word for sardines was exactly the same as the French—*sardines.* But to make sure I'd get some learning under my belt, he taught me to say the word for fish, instead. "Samak," I told Pouspous, holding up a tantalizing sardine, but not letting her have it.

I wanted her to learn the word, too. "Samak," I repeated sternly.

The cat blinked, as though she found Arabic befuddling. My father comforted her by slipping her another sardine, no strings attached.

I would never have as gentle a teacher, or as creative. Though he was known for being tough and stern, he was neither to me as we sat side by side, day after day, and practiced random words and phrases— whatever came into his line of vision: the cars floating along Malaka Nazli, the lively alleyway by the side of our house, the pretty girls who walked by and smiled our way, and any and all attributes connected to my cat, from the shades of her varicolored fur to her abiding passion for hard, salty *gebnah.*

It was all great sport for my brothers, who thought I was hopelessly inept and that my father was wasting his time. Arabic required a strong, guttural "h" sound. There were comical and even embarrassing results. The word for street is *harah,* pronounced with a harsh, emphatic "h." I seemed only able to say *kharah,* which is Arabic for "crap." I kept stumbling over the simplest words, the most basic sounds, and my brothers laughed and laughed.

My father emerged as my staunch defender, ordering them out of his room, then calmly resuming my Arabic lesson, with only Pouspous on hand.

"Mogrem," he shouted after them, "Ibn'el Kalb."

Those were two Arabic expressions I was delighted to memorize and say out loud. They meant: "Criminals, sons of dogs."

"Loulou, non," my father told me firmly and in French.

MY FATHER BEGAN AN international campaign to obtain a second opinion about his condition, and then a third, a fourth, and a fifth. He sought out specialists everywhere from England to France to Italy to America. He sent out letters courteously requesting the expert medical advice of the world-renowned doctors.

The specialists he selected were nothing if not exotic. They included a British knight; the director of an orthopedic clinic in Milan; and the renowned head of a medical institute in Bologna. He also hoped to connect with American specialists, and sent several copies of the large, ancient radiographies to my cousin Edouard, who had left shortly after his son's bris, and was now settled in San Diego, along with his wife and child.

Sir Reginal Son-Jones of London reaffirmed my dad's faith in the English with his immediate and gracious reply. He said he would like to examine him personally, and invited him to come immediately to England where "I will do all I can to help you." But he also noted his belief that surgery was inevitable: "Most certainly the nail will need to be removed."

My father enlisted the help of his nephew Salomone, who was living in Milan. Salomone had left Cairo in 1949 and returned to Italy, where he undertook an obsessive search to chronicle his parents' and younger sister's final days. He married an American woman—also from a prominent Aleppo Jewish family—but insisted on taking her to Milan to live. Still conscious of his debt to Zarifa and Leon for the years on Malaka Nazli, Salomone had gotten into the habit of saying prayers every night in his room, invoking my father's name. He didn't blink at the rather quixotic nature of the request. If Oncle Leon wanted him to tote large X-rays and medical reports to different European specialists, that is what he would do. He willingly accepted the packages containing the translucent black-and-white images showing a hipbone, a thigh bone, and broken pieces of a nail, along with an accompanying medical report.

The verdict was the same everywhere: my father almost surely needed another operation to repair the damage left by the other surgery, and to take out the broken pin that was causing pain and adding to the risk of complications. Yet physicians across Europe were optimistic that he would likely recover and walk normally again.

My dad, still traumatized by the memory of the hammer pounding the nail into his hip, as well as the months of anguished convalescence that followed, could not be persuaded to take the step, not even by the likes of Sir Reginald Son-Jones, or the equally eminent Professor Antonio Poli of the Instituto Gaetano Pini of Milan, or even by the distinguished Professor Scaglietti of the Instituto Ortopedico Rizzoli of Bologna.

From his perch by the window at Malaka Nazli, Dad continued his campaign. He scratched out letters to more noted experts at ever more distant medical outposts, hoping for the eighth, ninth, or tenth opinion that would be more to his liking: one that did not insist on another operation.

But he never made plans—actual travel plans—to go see any of these renowned physicians. Though the exodus of the Jews of Egypt was continuing, and my father now had the greatest impetus of all to leave—the possibility of obtaining more effective medical care than was available in Cairo—he still couldn't find it in himself to abandon Egypt. Instead, in between his cries and his prayers, he hoped that both the pain in his leg and the equally painful need to move would somehow vanish.

In San Diego, my cousin Edouard felt haunted by my father's letter. He detected desperation, even hopelessness—two qualities he had never seen in my father as he watched him from afar, a tall, confident, dreamlike figure in white striding through the streets of Cairo.

CHAPTER 9

The Lament of the Rose Petal Vendors

Loulou, éloigne-toi du balcon," my mother was crying; Come away from the balcony. I was at my favorite perch, looking down at the alleyway by the side of our house, Pouspous in my arms. Mom couldn't understand my fascination. She'd much rather I sit with my father at the window in his room: Wasn't watching pretty Malaka Nazli more fun than staring at some little side street filled with dirty urchins and peddlers? she asked reprovingly.

My father tried gentle suasion, coaxing me to return for more snacks and Arabic lessons. I'd hear him call out, "Loulou, Loulou." He, too, was sure that I'd prefer the more opulent street life of Malaka Nazli with its vast lanes of traffic and its constant flow of passersby.

I loved Malaka Nazli, but I enjoyed my alleyway even more.

Each morning, Pouspous and I watched as streams of merchants wandered by, pushing their wheelbarrows laden with fresh produce. They had been a fixture of our neighborhood since my grandmother Zarifa's day, hawking whatever they could pile into their carts—grapes,

figs, green beans, oranges, celery, okra, tomatoes, mountains of potatoes, or sweet luscious *mesh-mesh,* apricots.

The vendors were all impoverished Egyptian fellahin. The contents
of their carts were literally their only capital, the few piasters they made
off the sales of their fruits or produce their chief source of income.

And some couldn't even afford a wheelbarrow or a carriage. Instead,
they carried their inventory in large straw baskets perched on their
heads, and sang to announce their arrival and what they were selling. It
could be *samak*—fresh fish from the Nile—or grape leaves, *wara enab,*
neatly tied together in small green bunches, or parsley, or my favorite
of all, scented white rose petals.

Watching them turned into an obsession: I wondered how they maneuvered so gracefully with such a fragile, unwieldy burden. I couldn't
figure out how they prevented the petals from blowing in the breeze
and scattering all over the street. The problem consumed me until I
realized that they covered the petals with a large damp white cloth,
both to retain their freshness and to keep them from spilling or fluttering away.

A bushel of the blossoms cost only a couple of pennies. There was
an eager market for them among housewives who used them to make
rose petal jam, a time-honored concoction as popular now as it had
been in my grandmother's day, or when my aunts lived around us. Tante
Rebekah was renowned for *la confiture de roses,* a light flavorful jelly
prepared with a mixture of lemon and sugar and, of course, fresh rose
petals. Tante Leila loved making *maward*—rose water—an absolutely
essential ingredient for baking, which imbued any pastry or cookie with
the aroma of the flower.

It was said that the best petals were sold only one month of the year,
when they weren't even pink, when they were alabaster white.

In the season of the rose petals, the day began with the vendor wandering the streets, chanting "El ward, el ward, lel charigh ya ward"—
Roses, roses to make rose water—in a loud, piercing voice designed to
reach even those who lived on the top floor of a building.

If someone popped their head out of the window and signaled to the
vendor, he'd trudge upstairs with his overflowing basket and his little
scale. He wasn't selling the actual flowers, of course, only their petals.

At times, he offered the heads of the flower, minus the stems, leaves, and thorns, and they had to be taken apart at home, petal by delicate petal, to be distilled to their essence.

Once they'd made their sale, the rose petal vendors would vanish into the crowded alley. Yet long after they were gone, I still heard the echo of their chant, *"Elward, elward"*; Roses, roses.

Even when the roses were not in season, the alleyway still exerted its hold on me. My favorite spectacles were the ones I could watch unfolding from the balcony.

On any given day, Arab men in white flowing robes would set up ceremonial tents of one sort or another. Both weddings and funeral memorial services required elaborate preparations, and the alley below our house was a favorite setting for both. From the early morning, I'd watch the men setting up lamps and laying down large carpets. Pouspous was my steady companion. She knew instinctively that if she let me hold her, I would feed her little bits of her favorite foods—cheese, pieces of bread, any leftover chicken or sardines I could get my hands on from the kitchen—so she'd settle comfortably in my arms, on the wide balcony railing.

Together, the two of us would observe the goings-on, and I'd stroke her to keep her still. She seemed as riveted as I was by the activity below. Every couple of days or so, I'd watch, hypnotized, as groups of strange men set up a large tent, carted dozens of wooden folding chairs inside, and began to spread elaborate carpets on the ground.

Once a tent was up, it was a game for me to figure out what was taking place inside. If I heard music from an orchestra, or sounds of people clapping and women making the strange whooping sound known as *zaghlouta,* where they cluck their tongues and tap their hands across their mouth, I knew there was a joyful celebration under way, most likely a wedding or an engagement party.

But often, I'd see somber men trooping in and out and there was no music at all—only loud cries by the professional mourners with amber paint on their faces, which shattered the day and the night, and I knew that someone had died. Like black, amber was the color of death.

I wasn't allowed anywhere near a funeral, even that of a relative, let alone a stranger's. My parents put their foot down whenever there was a memorial service under way and my mother would order me to come immediately away from the balcony. It was as if I had to be kept away from death at any cost.

One morning, new neighbors moved into the street-level apartment in the building directly across the alleyway. It was a young Muslim couple, newlyweds who seemed very much in love. The groom was rather ordinary, and sullen, but the bride instantly became my friend. She was very pretty, with long dark hair, dark eyes, and a merry smile. She'd open the shutters each day, stand on the balcony, and wave and point to Pouspous. I'd wave back and hold the cat up in my arms. Finally, after some weeks of gesturing and shouted conversations from across the alley, she motioned to me to come around and visit.

I ran to my father and told him about my new grown-up companion: Could we go call on her?

My father seemed amused. Yes, of course, he said; he'd noticed our new neighbor from his window and would be delighted to take me to see her.

That afternoon, Leon and I ventured out to a nearby florist. We walked slowly, because he was still unsure of himself. We purchased an immense bouquet of roses, crimson and in full bloom. My father allowed me to carry them to the woman's apartment and watched, smiling, as I handed her the flowers.

My neighbor seemed surprised, almost flustered, by the size of the bouquet.

"What is your name?" she asked, as she took the flowers. In all those weeks, shouting across the alley, we had never introduced ourselves.

"Loulou," I replied.

It was the only word I had spoken in what had been, so far, a silent transaction. She told me she was moved that a child would give her such a lovely gift and invited us both to come in and sit in her living room. It was simply furnished, and much smaller than ours, but also modern, not weighed down with the ancient sofas and tables we had in our house.

Our neighbor was even prettier up close than she seemed from across the alleyway. Extremely slender, her hair flowing past her shoulders almost to her waist, she was constantly smiling and laughing, chatting mostly with my father in animated Arabic.

He was clearly beguiled.

I could always tell when my father approved of a woman, and he only approved of attractive women. He believed that beauty was the one quality essential to a woman—more important than wealth, social status, family background, and certainly education. He was adamant on this score. I could have asked him why he bothered to give me Arabic lessons and insisted I study.

Our hostess suddenly jumped up and began to dash about the house, in search of a vase large enough to accommodate all of our flowers. Much of the visit was spent watching her running in and out, darting from one room to another, her husband nowhere to be seen.

She seemed so agitated, and I assumed it was because of the roses: we had brought far too many. It didn't occur to me that my father had unnerved her. She kept eyeing the thick bouquet, holding it up against the small bud vases she had on hand, and putting it down again. At last, she decided she would divide the flowers among multiple containers. She enlisted my help as she searched for empty Coca-Cola bottles, pots, tall glasses—whatever she had lying around in the kitchen. She'd place a couple of stems in one container, four in the other, until all two dozen roses were finally steeped in water and on display, here and there, in different corners of the apartment.

In a trademark gesture of Egyptian hospitality, the young bride brought out a platter of pastries—*konafah,* a delicacy made of shredded wheat stuffed with walnuts, and *ghorayebah,* a soft, pale white cookie topped with a single green pistachio. She made a point of offering the desserts to me first before serving them to my father, in the same way that in the middle of their intense chat in Arabic, she would turn and try to engage me in the conversation, as if to underscore that I was her principal guest, the one who had, after all, brought roses.

When she had a child, she informed me, she would name her Loulou, after me. "Even if it is a boy, he'll still be called 'Loulou.' " She laughed. "Then you will have someone to play with besides that cat," my neighbor added with a wink.

A couple of months later, I heard a loud cry from the courtyard. I ran to the balcony. Somber old men began setting up a tent, spreading dark crimson rugs on the ground. I could hear wails throughout the day and night; the professional mourners were out in full force. My mother commanded me to leave the window at once.

"Loulou," she said sternly, "Éloigne-toi de la fenêtre tout de suite. Il n'y a rien à voir." There was nothing to see, she repeated.

This time I obeyed her, and walked to the corner of the dining room, sat on the Persian rug, and began to cry out loud like the mourners. I cried and cried. Pouspous joined me, and in a supremely tender gesture that endeared her to me forever, she began to lick away my tears, her small tongue darting across my cheek.

It was my friend, the young bride, who had died. The shutters across the alleyway were never open again. I never understood what had happened, and no one in my family would tell me, neither my mother, who surely knew more than she was letting on, or my father, who had befriended her that day we'd brought her roses. Her passing emerged as one of the great mysteries of my childhood in Cairo and one that I could never stop mulling over, especially in the months that followed when I suddenly found myself struck by a bewildering malady of my own. My own brush with illness would make me obsess even more over the loss of my lovely companion. I yearned for her supremely reassuring presence, smiling and waving from across the alley.

CHAPTER 10

The Cure for Cat Scratch Fever

Shortly before my sixth birthday, I developed a series of mysterious ailments that baffled the entire Cairo medical community.

It all began with a low-grade fever that refused to go away, along with a strange swelling in my thigh. My symptoms multiplied to include a rash, any number of aches and pains, and finally, an overwhelming sense of lethargy that seemed entirely at odds with the playful child I had been.

Friends and well-wishers inundated us with advice, from crude homemade remedies to the latest scientific therapies.

Mom's favorite seamstress prescribed a poultice that was to be applied on my leg at least once a night. Before I went to bed, a smelly concoction of warm dough, mustard, vinegar, and herbs was placed on me for an hour or two, or until I got too fidgety and yanked it off. A series of doctors descended on Malaka Nazli to administer injections, as well as prescribing an array of strange pills and potions. My sister Suzette, who dreamed of medical school but settled for a job teaching kindergarten at the Lycée Français de Bab-el-Louk, suggested finding a top specialist, someone with offices *en ville*—preferably a European—as no one my parents had brought in met with her approval. Our Sudanese porter Abdo advised us simply to pray.

Nothing worked.

My father, by now a veteran of long, complicated ailments, decided to take matters into his own hands. Together, we made the rounds of Cairo's top physicians, going from one to the other, one to the other, in search of an answer to what ailed me. Most of the doctors had been treating him for years, and knew him on a first-name basis. Ever since his fall shortly after I was born, my father had endured recurrent, painful flare-ups of his hip and leg. He'd never had the broken pin removed, and he limped noticeably.

We made an odd pair as we toured the specialists' offices—a tall, distinguished silver-haired gentleman whose right leg dragged, walking hand in hand with a diminutive girl with long dark hair and dark eyes, who shuffled ever so slightly on her left leg.

My dad seemed at home with these elite physicians, whose offices were in the most upscale section of the city. Could you please examine my child? he asked each of them politely, almost obsequiously, as if they'd be doing him a favor by adding his slip of a daughter to their practice. As if he weren't going to reward them with wads of large bills from his wallet.

The Cairene medical establishment was frankly bewildered. The blood tests came back negative. X-rays taken of every part of my body revealed nothing out of the ordinary. The syrups and pills and injections seemed to have no lasting effect. The swelling in my leg refused to subside, my fever kept returning, and I had also developed a slight rash. I felt so tired that I didn't want to play anymore, save with Pouspous. One doctor after another was unable to say what was wrong, beyond insisting that whatever ailed me was undoubtedly benign.

At last, my sister located a physician known reverentially as "Le Professeur." Suzette was certain the doctors my parents were consulting on my behalf were leading us down a primrose path. She was always at war with my father, persuaded he was letting me down, and my care had become the latest pretext for her rage.

Le Professeur was a man of such renown that few ever bothered to use his name. He was revered by his patients, who included affluent Jews and Arabs, as well as whatever was left of the European community. His strange habit of always wearing white cotton gloves—to hide a skin condition, probably eczema—only added to his mystique, and to my fear of him.

After closely examining me, in particular, the lump, he ventured a diagnosis: I had contracted a case of Cat Scratch Fever, *la Maladie des Griffes du Chat*. My father, no stranger to medical jargon, looked startled at the improbable-sounding verdict. What on earth was the *Maladie des Griffes du Chat*, he asked, and why had it taken so long to figure out what was wrong with me?

The Professor grimly described an obscure disease, first isolated by a French scientist in the 1800s. Even now, he said, Cat Scratch Fever was known to only a handful of practitioners, and no doctors save the few who had studied in Paris and other great European medical centers were even vaguely familiar with the disease that was apparently attacking my entire immune system. The fever was thought to be caused by a scratch or bite from a cat that unleashed a painful inflammation of the lymph nodes. Children were the most vulnerable. No doubt, my cat had scratched me near my thigh, resulting in the unsightly cyst he called *un ganglion*.

I didn't understand much of what the doctor said—only that he seemed to be blaming my illness on Pouspous, whom I held in my arms much of the day and who ate with us at our dining room table. How could my cherished calico be to blame for the injections and blood drawings and legions of men in white coats who were constantly poking me? The uncertainty, the fear, the possibility that I was dangerously ill—it was all the cat's fault.

Having rendered a diagnosis, the Professor seemed far less certain of the cure. Cat Scratch Fever was still a medical mystery, he told us, and no one really understood it. While antibiotics were thought to help, there were no definitive remedies. The Professor could only express the hope that my Cat Scratch Fever would go away by itself.

Children, he told my father as he patted my head, can heal almost by magic. They can seem deathly ill and then, overnight, be well again.

"Dieu est grand," my father exclaimed. He said that whenever he felt hopeful or frightened.

The Professor nodded in agreement, but as if to hedge his bets and ours, he made us an offer. If I came back to his office once or twice a month, he would monitor me and determine whether I was showing any signs of improvement. He attempted a smile my way. I didn't smile back; I kept looking at his white gloves.

They made me anxious; would he ever take them off?

Vowing to bring me back promptly in two weeks, my father took me by the hand and walked with me outside to the street where we hailed a taxi.

He reached into his pockets and pulled out several pieces of candy in pretty silver wrappers. "Loulou, prend," he said, handing me some bonbons, my reward, I suppose, for having endured the Professor's relentless examination. My father never left the house without a supply of candy as well as the Lucky Strike cigarettes he kept in a silver case. Though he never lit up, he loved to give out the popular American cigarettes to friends and business associates.

He asked the driver to take us to my favorite hangout in all of Cairo: Groppi's. It was a surefire way, he knew, to cheer me up—as simple as letting me order a cup of *pêche Melba*. Sitting there with Dad, amid the splendor of Groppi's pebbled garden, eating spoonful after spoonful of the delectable peach ice cream that came in a tall fluted glass, topped with whipped cream, I almost managed to forget the white-gloved doctor.

The legendary patisserie was struggling to survive in the postcolonial world, even as many of its clients left the country. It still maintained two locations, its grand flagship store on Suleiman Pasha, and the one I liked even more, on Adly Pasha, because it had outdoor tables set within a pebbled garden. This symbol of foreign decadence had managed to forge a peace with the new revolutionary rulers, and it was whispered that Nasser and Sadat stopped by on occasion, if discreetly. The remaining expatriate community, the smattering of Jews and French and Italians who had somehow hung on, kept Groppi's as their unofficial headquarters.

My father could tell that I hadn't liked the Professor, but he was still

*Groppi's
of Cairo.*

in a cheerful mood. Merely having a name for what ailed me was progress. For so many months, my family had lived in a state of uncertainty,
wondering what was wrong with me. I had even overheard my older
sister use the word *tumor* to describe the lump that refused to go
away.

I had no idea what a "tumor" was, but I guessed from the way Suzette said it that it was bad. Very bad.

When we came home, my mother rushed out to ask what the latest
specialist had said. Almost immediately, the order came down: I was to
stay as far away as possible from Pouspous. "Tout ça, c'est à cause du
chat," my mother said crossly; It's all because of the cat. At last she had
an object for her rage, someone she could blame for my strange illness.
I had the distinct sense that if she could, she'd simply boot out Pouspous and forbid me to ever play with her again. She did make me promise to stop embracing the cat or carrying her around in my arms, as I
loved to do.

———

I TURNED SIX IN the fall of 1962, and my family was in the throes of deciding whether to abandon a country they loved deeply but which no longer wanted them, or whether to tempt fate as well as the authorities by trying to stay.

My father, who couldn't fathom life outside of Egypt, opted as always for holding on. His health was so fragile, he couldn't envision a lengthy voyage. Besides, his business was here. His synagogues were here. And though nearly every single member of his extended family—brothers and sisters, nephews and nieces and cousins too numerous to mention—had left, he still had friends and acquaintances in Cairo.

But it was getting harder and harder for Dad to resist the pressure: Jews were leaving in droves.

The convulsive departures were under way throughout the Middle East. Countries where Jews had lived harmoniously with their Arab neighbors for generations found their situations untenable. One after the other, Jewish communities in Libya, Algeria, Yemen, Iraq, Tunisia, Morocco, Lebanon, and, of course, Egypt dispersed. They left behind magnificent synagogues, schools, hospitals, and a way of life that had been, in many of these countries, often blissfully free of intolerance. World War II was fresh in everyone's mind, as were the mistakes the German and European Jews had made in staying where they weren't wanted. The world, still reckoning with the aftermath of the Holocaust, was determined not to risk another slaughter of Jews. Country after country opened its doors, and from the late 1940s through the early 1960s, nearly one million Oriental Jews scattered to the four winds. They left for Israel and America, as well as fanning out to Italy, England, Spain, France, even Australia and distant corners of Latin America.

Alas, what no one could stop was the cultural Holocaust—the hundreds of synagogues shuttered for lack of attendance, the cemeteries looted of their headstones, the flourishing, Jewish-owned shops abandoned by their owners, the schools suddenly bereft of any students.

We were witnessing the end of a way of life that many would look back on with a mixture of bitterness and longing.

Some, like Dad, tried to stay put, unable to accept the finality of the situation.

Our lives seemed to go on as before, but beneath the veneer of normalcy there was a feverish uncertainty. We didn't know whether we'd be around another month or another year or another ten years. I was supposed to start second grade at the Lycée Français de Bab-el-Louk, but my father still hadn't paid the tuition. It wasn't that he couldn't afford it. He still had means to pay for the best doctors, the best restaurants, and the best schools for his children. But it was such a tenuous time, and he was feeling squeezed by the Nasser regime. Most painful to my father was the evisceration of his beloved stock market. Nasser had nationalized one company after another, which made many of Dad's holdings worthless.

"On a tout perdu—on n'a plus rien," he told my mother one afternoon; We have lost everything—we have nothing left. He was close to tears, especially since my mom had always disapproved of his passion for investing, and had felt that stocks were a risky proposition. Tears were very unusual for my father. My sister Suzette had a twenty-pound note on her; she went over and offered it to him. He refused.

But he held off paying my tuition: did it make sense for me to begin a school year I might not be able to finish? As a result, the headmistress, quite sensibly, wouldn't let me enroll. This distressed my mom to no end. Without telling Dad, my mother took me across the street for an interview at the Convent of the Sacred Hearts. One of the most respected schools in all of Egypt, it had, on occasion, admitted Jews, including my older sister.

When my father found out what she had done, he was furious. He had always felt Sacred Hearts paved the way for Suzette's rebellion; he wasn't taking chances with me.

"Jamais de la vie," he said. Never.

But my mother, who rarely asserted herself, was adamant. If Sacred Hearts was out of the question, then he would have to go at once—at once—to the Lycée Français and settle the debt to allow me to attend classes with the other children. The next day, I found myself once again enrolled at the Lycée de Bab-el-Louk, walking round and round the vast courtyard with my friends.

My illness became a diversion from the larger pressures we faced. The burden of caring for a sick child and an ailing, much older husband, combined with the need to make a firm decision about whether and when to leave, was taking its toll on my mother. She was teary-eyed and fretful, wondering what would become of us, most especially "pauvre Loulou." Since the onset of the fever, that is all she ever called me.

Fueling the pressure was Suzette's insistence that we abandon Egypt immediately. Her own friends had been leaving one by one, and these days, she had almost no one left to join her for a movie or an afternoon of shopping, other than the stray Muslim girls she had befriended. She was anxious for the next chapter of her life to begin. Because she was less emotionally vested, she also saw the dangers of staying put in Egypt more clearly than my father. But she couldn't seem to persuade him, and her words fell on deaf ears.

It was as if my dad believed the turmoil was only a passing phase. The political situation would settle down as it had so many times before. The fever in the streets would subside, and he'd be able to stay exactly where he was, enjoying the street life from his ground-floor window on Malaka Nazli.

There was a truce only when the entire family came together to discuss what to do about me. They gathered in the living room, and talked as if I weren't there, as if I couldn't possibly understand what they were saying. I overheard them analyzing my illness and its dismal prognosis.

Though my contacts with Pouspous had been cut back, the Cat Scratch Fever remained. I was still able to go to school, but I didn't feel well. My symptoms receded and flared, receded and flared. My left leg was still swollen, the lump was still there, as large and hard and menacing as ever. I ran a fever that aspirin couldn't control. I went regularly to see the Professor, or he came to Malaka Nazli. I dreaded the examination, especially his white gloves. Yet he was gentle enough as he searched for a sign that the strange swelling in my leg was going down, or that the fever was abating. After a couple of months, he declared himself at a loss.

In what amounted to an admission of defeat, he suggested that perhaps I would be better off if seen by specialists in France or America.

My father seemed shaken. The doctor who had offered a powerful incentive for us to stay now cited compelling grounds for us to go. We came home and relayed to the rest of the family the disheartening news. Cairo's top specialist was urging us to seek help elsewhere. There were no more answers left in Egypt.

Over the years my father had been ill, he was often told that doctors abroad were likely to have more options for him than in Cairo, which despite its pretensions to being a simulacra of Paris was still in many ways a Third World city. After his fall in 1958, Dad had been in perpetual pain. But as he improved, and learned to control the pain, in large part because of patient ministering by legions of Cairo's leading physicians, the incentive to seek out other opinions waned, and with it, the incentive to leave.

With my vexing malady, my father once again heard the siren song of the West.

Both he and my mother longed to find the mythical *bon docteur* who would be more knowledgeable than his provincial Cairo colleagues. Surely, in Paris or Milan or New York, there would be a physician who could make me well again.

AS HE WAVERED, MY father was reminded of the advice of our porter: it was time to pray.

For that, my father favored the Kuttab, the little shul around the corner. Every morning for years, my dad would walk to the homelike structure, bringing supplies of coffee, tea, and sugar that a servant would prepare for the worshippers. He'd linger till the late morning, returning at the end of the day for evening services. There were times he stayed all night praying and bantering. The Kuttab was as much a social as a religious affair, almost like a private club where a group of prosperous Levantine businessmen who had known each other for decades put aside their worries and concentrated on praying as well as gossiping.

Though cozy and intimate, it was also oddly snobbish and deeply fashion-conscious.

My dad and the other synagogue elders made it a point to wear

white, all white, to services each week. Indeed, they dressed as meticulously as they had in the 1940s, when standards were set by the British
colonial officers. Even now, with the officers gone, and Egypt in its
shabby and dejected postcolonial phase, the men of the Kuttab made
sure that every stitch they had on was a dazzling shade of white, down
to their shoes, though some favored white wingtips with a brown trim.

I'd often accompany my dad to Saturday-morning services, and
while women were relegated to their own section, I was allowed to sit
with him and the other men, which I considered a wonderful honor,
and one that I tried to earn. Instead of playing with the other children,
I'd take a Hebrew prayer book I didn't know how to read and pretend
to follow the liturgy. When I was restless, I'd wander outside, where
groves of jasmine were in bloom. Their scent was so alluring, I'd pluck
the delicate white flowers and make small garlands to bring back to the
sanctuary and walk around, handing out blossoms to all the men to
wear in their lapels.

My favorite moment was when everyone stood to receive the priestly
blessing. This solemn affair called for any man who was a *cohen,* a de-
scendant of the ancient order of High Priests, to stand in front of the
sanctuary and bless the congregation. All the male worshippers were
expected to rise, drape the prayer shawl over their heads, and cast their
eyes downward to the floor. My father insisted on having me at his side
as he lifted the shawl over our heads to construct a makeshift tent. He
would place his hand over my head, as if to confer an extra measure of
priestly benediction.

I never felt so safe as those moments beneath the white prayer
shawl. There was nothing to fear, not even the ravages of Cat Scratch
Fever.

Still, because my illness continued to take its toll, my father decided
to expand his repertoire of synagogues. His favorite, the Congregation
of Love and Friendship, had closed a few years earlier after most of its
members left for America. But there were still at least half a dozen
functioning synagogues in our neighborhood, ranging from the intimate Kuttab to the more stately Temple Hanan, with its vaulted ceilings and spacious courtyard. Off we went one morning to Temple
Hanan, the membership of which had dwindled, where he asked the

rabbi to make a special prayer for me. Placing his hand over my head, the rabbi chanted to God to deliver me from Cat Scratch Fever. He then reached for some fragrant rose water in a silver container, and drizzled it all over my face and arms.

At my father's urging, my mother and I began making the rounds of Cairo holy sites, where miracles were known to occur.

We began by visiting the Gates of Heaven, the most important synagogue in all of Egypt, built in the nineteenth century. Legend had it that its wealthy benefactors had thrown precious gold and silver coins into the foundation for good luck. It was at the Gates of Heaven where my parents were married in the spring of 1943, standing at an altar strewn with white roses.

My mother took me by the hand up the same marble steps she had climbed as a twenty-year-old bride. The precious Torah scrolls were kept behind a velvet curtain, and she lifted me in her arms so I could kiss the curtain, which itself was holy. "It will bring you luck," she whispered.

We journeyed next to Ben Ezra, the synagogue where several of the biblical prophets were known to have lingered. Jeremiah, the prophet of lamentation, was buried beneath the temple's stone foundation, while Elijah, who was so pure that legend had it that God couldn't bear to let him die, was said to stop here in between his sojourns performing good deeds across the earth. Ben Ezra was located in Old Cairo, the most ancient part of the city. The morning we went it was deserted, its pews devoid of worshippers.

As we walked out, my mother left behind a small gift: a jar of white sugar and a bottle of fragrant orange water. They were treats for Elijah, she explained, so he would look kindly on me and come rid me of my Cat Scratch Fever.

On the other side of the building was the famed Cairo Geniza, an attic where leftover parchments and pieces of prayer books, some dating back hundreds of years, had been "buried," because it was forbidden to throw away so much as a scrap of paper with God's name on it. The Geniza's treasures had been carted off years earlier to Cambridge for scholars to study. But my mother still motioned in the direction of the attic and told me to pray, because even the abandoned burial site was holy and could effect miracles.

At the end of each of these pilgrimages, my mother conferred with my father. He was engaged in his own efforts to heal me, of course, though they were more discreet. Every morning, he led special prayers either at the Kuttab or Temple Hanan on my behalf. Every night, he lit a glass filled with oil and a floating wick, as he prayed for my recovery.

We saved the most important journey for last. It was to Rav Moshe, the Temple of the Great Miracles, the synagogue of Maimonides, the great healer himself. The ancient little building was located in the heart of the dusty ghetto known as Haret-el Yahood, literally "the street of the Jews." It was a neighborhood where only the poorest members of the Jewish community, those whom time and fortune had left behind, still lived.

None of us ever ventured there. As Cairo Jews prospered, they moved as far away as possible from the Jewish Quarter, renting fashionable apartments downtown or in my family's area, up and down airy Malaka Nazli Street. They had almost no dealings with those who still lived in the ghetto.

Except when calamity hit.

Then, even the most elegant Cairenes voyaged to Rav Moshe, the Temple of the Great Miracles.

Legend had it that Maimonides, the renowned twelfth-century Talmudist and physician who believed equally in the power of medicine and magic, had performed several of his renowned feats of healing deep inside the temple walls. No one knew exactly what these acts were, but the image of Maimonides practicing both his faith and science within this one small edifice was enough to draw visitors from all over the Levant. The Temple of the Miracles became a kind of Jewish Lourdes. Mothers brought their sick and crippled infants, children accompanied dying relatives, widows and widowers hobbled in by themselves to seek help.

I followed my mother down a short flight of steps to the cool, dark, cavelike structure that lay beneath the main sanctuary. Small alcoves had been carved out from the gray rock and transformed into sleeping areas, with thin mattresses and pillows, sheets, and blankets. The makeshift beds were so low it was almost like sleeping on the ground. The room was almost pitch-black, with only a few faint rays of light coming through the small square windows, which didn't even have panes. I could see the outlines of people lying within the alcoves. Some were elderly and lay completely still, but there were also mothers huddled with their crying babies.

An old man appeared, seemingly out of nowhere, and handed us a glass filled with oil. It was the oil of Maimonides, the Great Healer, the Maker of Miracles, he said, as he pointed to a spot in the floor. He whispered that somewhere there, beneath us, Maimonides' finger lay buried. The man instructed my mother to rub drops of the holy oil all over me to make me well. Was there a special problem? he asked, turning to look at me.

My mother pointed to my left leg. "Elle a la Maladie des Griffes du Chat," she replied sadly.

The man nodded, as if a thousand cases similar to mine had come his way, and Cat Scratch Fever was a common ailment he saw every day. He urged her to pour extra drops onto the afflicted leg. Then he pulled out a collection box. Without saying another word, my mother opened her purse and pulled out coins and bills and stuffed them in the small box. The man bowed and disappeared into the darkness of the catacomb.

I clutched my mother's hand as she searched for an empty alcove. When she located one in a corner, she told me to lie down and go to sleep. But first she did exactly as the man had said, rubbing oil over my leg as she said a prayer to Maimonides.

"Make us a miracle now," she whispered.

She looked around furtively, as if expecting the Great Healer himself to step out of the darkness.

I must have seemed frightened, because she promised she wouldn't leave me, that she'd sit by my side—all night if need be. What was important, she said, was that I sleep. Maimonides wouldn't come while I lay awake. He couldn't perform his miracles if I watched.

I still couldn't close my eyes, so my mother resorted to what she had done with me at home on a thousand nights; she told me a bedtime story. Sitting on the edge of my hard mattress, she began with the familiar soothing refrain, "Il était une fois . . ."

Once upon a time, there was a fabulously rich woman—a friend of the family—who had visited every doctor in Cairo because of her finger. It had become red and swollen and seriously infected and was now in danger of developing gangrene. All the doctors near fashionable rue Kasr-El-Nil insisted the finger needed to be removed at once. On the eve of surgery, the woman made the journey from her home in Maadi, the wealthiest neighborhood in all of Cairo, to the temple in the poorest neighborhood in all of Cairo. She came alone, without her usual retinue of servants, and lay down in an alcove exactly like mine, my mother said, with nothing but a thin blanket and a frayed old pillow.

When she woke up the next morning, her finger was clear of infection. Maimonides had come and cured her so completely that she didn't need an operation. And the same would happen to me, my mother vowed. By the morning light, I would be free of my Cat Scratch Fever.

That night, I could hear people from other alcoves calling for the long-dead rabbi and healer to come save them. At last, I managed to close my eyes. Though my mom never left my side, I was still afraid— afraid of the cave, afraid of the moans all around me, afraid that I would be left here forever in the dark, afraid most of all of coming face-to-face with Maimonides.

I wanted to be home with Pouspous in my arms, on Malaka Nazli Street.

I must have drifted off, because when I woke up, it was already morning. Little bits of sunlight were shining through the small window, and I could see that my mother, true to her word, had stayed the entire night in her awkward position, neither comfortably seated nor lying down.

The room felt less oppressive, and there was life and movement as people began to shuffle out. I saw a woman carrying her sick child, and a young girl helping an aged man, probably her father, perhaps even her grandfather, up the stairs. No one seemed to be moaning anymore.

Had there been a miracle?

The man who had given us the oil reappeared, this time with a small washbasin filled with water. My mother could help me rinse off the holy oil, he said, then we were free to leave. She smiled as she applied the cool water all over me, certain we had found a cure for my mysterious malady.

CHAPTER 11

The Wayward Daughter

No one on Malaka Nazli clashed with Dad as frequently, or bitterly, or hopelessly, as my older sister. It was as if, even at eighteen, she was still seething over the initial injury, or perceived injury—the fact that he had named her Zarifa, and that even now he refused to use her pretty adopted moniker, Suzette.

The more his autocratic streak asserted itself, the more she bristled and rebelled. He loved religion, and embraced all the rituals and traditions of being Jewish. She hated Judaism and began breaking one by one with its tenets and traditions. He liked the old-world feel of our Ghamra neighborhood and our ground-floor apartment; she longed to flee Malaka Nazli for the swankier high-rises downtown. He hoped she'd settle down and get married. She despaired that her schooling was over, and still hoped to go to college. He wanted her to find a nice boy from one of the families we knew; she taunted him and all of us with her longing for "un blond aux yeux bleus."

The city was teeming with foreigners, though not the British and French and Belgians of days gone by. Nasser made no secret of his

Suzette's cinema club membership card.

friendship with the Soviet Union, and ever since the Suez crisis, the bonds had intensified, so that there was a visible Russian and Communist presence throughout the Egyptian capital, along with representatives from Eastern Europe—East Germans, Czechs, and Yugoslavs.

Many were "blond aux yeux bleus." Handsome, exotic, off-limits, they were the ones my sister fancied. Suzette flouted my dad's authority, at times outrageously. She stayed out later and later, as her circle of friends expanded. None were Jewish, of course, and that was the point.

Mom was caught in the middle, helpless to modulate her strong-willed husband or subdue her equally stubborn daughter. I watched, puzzled, unable to understand the velocity of the fights. I adored my dad, and my relationship with him was intensely peaceable. The bond that had been cemented in those months after his accident, when he and I were thrown together, a man in his late fifties caring for an infant, had only strengthened. Besides, we were like-minded, and our temperaments were similar. Whatever he loved, I loved—shul, Groppi's, Pouspous, the name Loulou, and of course Malaka Nazli.

They were heading toward an epic clash, and yet we were all unprepared when it happened.

The call came in the middle of the night. It was the police, and they were insisting on speaking to my father. "Your daughter is under arrest," they said. "We suggest that you come at once to police headquarters."

My teenage sister had been picked up and thrown in jail, but no one seemed to know why. The rumors were swirling.

Suzette had been partying with Russian spies. She had gone dancing with Norwegian sailors. She had joined Czech diplomats at a swank hotel, and the hotel had been raided by cops, vigilant under the new dictatorship.

Then there was the theory that my teenage sister was simply an innocent victim, a pawn in the growing political upheaval. In 1962 Cairo, arrests had become almost commonplace. The ruthless hand of the Nasser dictatorship was everywhere, and any illusions that the new regime would be more enlightened than the king's had disappeared.

Authorities were eager to strike fear in the hearts of all those derided as "foreigners," any vestige of the old colonial crowd, Jews, anyone, in short, who had overstayed his welcome and was still in Egypt when he ought to have left.

We had all come to fear the knock on the door in the middle of the night.

"Why didn't we leave?" my mother asked my father that night, and every night thereafter. "We should have gone years ago." She couldn't stop crying.

She offered to accompany my father to the police station. He shook his head no. Instead he contacted a friend of the family, a businessman named Monsieur Gattegno who lived down the street, and asked him to come with him. Though he tended to maintain a composed demeanor even during great crises, Dad appeared more agitated that night than I had ever seen him before.

He kept pounding his fist on the wall, and shouting that we were ruined.

He dressed slowly, as if he were having trouble moving. He was in his pajamas when the call came, and now, merely putting on a shirt and trousers seemed to pose a challenge. He threw on an old jacket that he rarely wore, a faded tie, and a careworn straw hat, and then called out to our porter, who was fast asleep in the apartment he occupied in the basement, to go please find him a taxi. He grabbed the wooden cane propped near the door, which he almost never used anymore, and slammed the door behind him.

With each passing hour, my mother became more and more distraught. My brothers were stomping around, looking bewildered. Neither got along especially well with my sister, yet the unfolding drama seemed less her comeuppance than some apocalyptic event destined to engulf the entire family.

I couldn't sleep because of all the commotion. I'd wander from the bedroom to the living room, where everyone had gathered, waiting for news. I held Pouspous firmly in my arms—my mom was so distracted that she didn't even order me to put the cat down. My fever had abated since the night at Maimonides' temple, either because of the mystic's intervention or a new antibiotic that I'd been prescribed by the Professor.

Every few minutes the phone would ring, and I'd overhear snatches of conversation. My mother would grab the receiver away from my brothers, and she'd cry out in disbelief, verging on hysteria: "Des espions," or "Un suédois." As the night wore on, there were more phone calls, and I heard a tangle of references to Danes, East Germans, Cubans, Poles, Norwegians, as well as Russians. There were mentions of soldiers, sailors, diplomats, even senior government officials. With each new rumor and revelation, my mother would dissolve in tears, and I could only clutch Pouspous more tightly, as if my cat would protect me from the mayhem.

Downtown at the police station, my father affected a remarkable transformation that had begun with his taking the cane. The proud and haughty would-be British officer known as the Captain was gone. Instead, what the policemen and militia members saw was a stooped, mild-mannered older gentleman with teary green eyes, leaning heavily on his brown wooden walking stick, in the company of another dignified elderly gentleman. He bowed to the officers and asked them for help in securing his daughter's release. As each gendarme disclosed a few more facts about the events of the previous night, my father deftly slipped them notes from his wallet. As the hours passed, so that dawn was almost breaking, my father continued to go from one officer to another.

Each time, he reached into his pocket and pulled out several flavorful bonbons—the same he routinely offered me or the neighborhood children or an attractive woman. The officers seemed delighted to accept the candy. He continued discreetly handing out the notes and candy, up the chain of command.

At last, the police captain came out. He was from the old school, a veteran. As he and my father talked, they found that they had a great deal in common, including the sense that modern youth was hopelessly lost, and that keeping a daughter in line could be trying for the most diligent of fathers. It was so much easier to raise a son, the chief sighed, and my father sighed along with him.

My father walked a tightrope—admitting Suzette had strayed, even as he tried to downplay the incident as the action of a naive young girl who, like so many of her peers in this new and faithless generation, had lost her way and gotten in over her head. He offered the chief a cigarette from the silver holder that was, as usual, filled with Lucky Strikes. The cop accepted an American cigarette, and as he lit up, he nodded thoughtfully that the incident had likely been an unfortunate mistake.

That was when my father began to cry. There, in the early-morning hours, in the middle of Cairo's main police station, as tears streamed down his cheeks, he had to be helped to a chair to steady himself, while the chief offered his own white cotton handkerchief to Monsieur Leon and tried to comfort him over his wayward daughter.

Inside, my sister and her friend Doris, a teacher at the lycée, were distraught about their night in jail. They still didn't understand why they had been arrested and, even more urgently, why their parents hadn't yet come to rescue them. They were tired, hungry, and bewildered. Their privileged upbringings hadn't prepared them for this stint with Cairo's underworld, the small-time prostitutes and female petty criminals with whom they shared a cell.

At last, my sister and her friend were released. Suzette was confronted by a torrent of insults from my father, who was both genuinely furious and anxious to show the police that he in no way condoned his daughter's conduct. Enough cash had been distributed, enough words of contrition uttered, that my sister was free to go.

She had been in jail the entire night.

The two finally came home; they walked in together, the patriarch and his eldest, wayward daughter.

It was striking how much they resembled each other. Unlike my mother, who was petite and fine-boned, my sister was tall and striking, a female version of my dad. She had his aquiline nose and full mouth, and even the shape of their eyes was similar, though his were a vivid shade of green and hers were coffee brown.

Most of all, she shared his strong-mindedness. Though she never would have admitted it, she was every bit as imperious and domineering as he, as much a creature of Aleppo, though she'd been born and bred in Cairo and modeled herself after the Europeans. She looked strangely defiant as she walked in; there was even a slight smile on her face.

My father hadn't said a word to her since springing her from prison.

But once inside the house, he exploded.

"You have ruined us," he kept saying. My sister said nothing; she acted as if she were impervious to his words.

Then he slapped her. In the middle of the living room, in front of my mother and brothers, while I hid in a corner with Pouspous. My sister retreated to her room, hating him more than ever.

That evening, Dad left the house and made his way to the Kuttab. Prayer services were long over; no one was there, but it didn't matter. Seated in his favorite chair, at the front of the small sanctuary, my father began to pray. He stayed until dawn, unable, unwilling to come home.

In the days that followed, I picked up only a few frightening words—"spies," "Russians."

My sister had crossed an invisible Maginot Line in a city that was, despite its cosmopolitan image, increasingly reverting back to its Islamic roots. Additionally, at a time when the Nasser regime had made it clear it wanted all Jews out of the country, it was quite plausible the government would have targeted a young Jewish woman from a good family, knowing that such a move would put pressure on her parents. My father was certainly attuned to the drastic changes that made the Jews of Egypt fearful and on edge. He grasped better than anyone the irrational, despotic nature of the new regime, so that he should have considered the fact that a couple of attractive young women who looked a bit too Western, who were behaving a bit too freely, made choice targets.

But in his anger and fear, that didn't matter. When it came to raising a daughter, he followed the ways of Old Aleppo, so that his moral code was as strict and unbending as that of his Muslim neighbors. He was furious that my sister had acted so imprudently, and could only ponder the shame of it all, and not that she may have been perfectly innocent by modern standards, or innocent enough. That would have made Suzette a double victim—of the strict, tyrannical Egyptian authorities, and of our strict, tyrannical father.

I tried several times to ask my sister directly what had happened, but she refused to say a word about it. It was as if she had made a vow never to revisit the incident.

Decades later, she would recall for me a balmy Egyptian evening. She and her friend Doris had gone downtown. As they strolled up and down the street in their prettiest dresses, sipping Coca-Cola and enjoying the weather, they had run into two dashing young men in uniform.

The men weren't Russian spies, my sister laughed.

They weren't even Russian.

They were actually Swedish, members of the United Nations peacekeeping delegation assigned to patrol the Middle East. They were handsome, friendly, and belonged to the Blue Berets, the official UN peacekeeping force.

"I can even remember their names." Suzette giggled, sounding more like the eighteen-year-old she once had been than the sober sixty-year-old woman she was when we spoke. "They were called Lars and Sven, and they had on these darling pale blue berets."

The four had gone off to a popular restaurant. "We were only having pizza," my sister insisted, underscoring the innocence of the evening. Then, suddenly, out of nowhere, several policemen and militiamen had swooped down and arrested her and Doris. They had let the UN representatives go, and taken only the two young girls. Handcuffed, forced to ride in the back of an open truck, they were taken to the city jail and placed behind bars.

"We were with the thieves and prostitutes and pickpockets—all the female criminals," my sister said, laughing, as if it had all been a lark.

César's recollection placed the events in a far less frivolous, less innocent cast. The arrest, he recalled, had taken place in a hotel, not a pizzeria, and perhaps inside a hotel room, where authorities had been alerted about two young women who had been seen entering with a couple of foreigners. In a paranoid era of a paranoid regime, my sister and her friend had pushed the envelope, and paid the price.

But on what charge? Surely, I persisted, there was a method to the madness of the Egyptian authorities, and even back then, they wouldn't simply have arrested someone without provocation. Under what ostensible grounds had Suzette and Doris been thrown in jail? Could it have all been trumped up? It had been entirely trumped up, she suggested— a bewildering, terrifying episode that had taken place in bewildering, terrifying times.

In that light bantering tone my sister had assumed throughout the conversation, she furnished the answer that had eluded me all these years, since the night at the police station, when my mother had stayed up crying on the sofa, and my brothers had paced self-importantly, and I had held Pouspous tightly, and the family had decided to leave Egypt—had decided that it was useless to try to stay because my father himself had deemed our lives there to be effectively over, that there was no point in even trying to hold on anymore.

The answer came in a single word: "Prostitution."

Within weeks of my sister's arrest, my father had secured the proper papers the family needed to leave Cairo.

CHAPTER 12

Last Call at the Dark Bar

As a little girl, I believed there was only one bar in all of Cairo, perhaps in the whole world, and that was the bar in the Nile Hilton. Though most people ended their day at a bar, it was often our first stop when my father and I set out in the morning, the place where we went to get our day going.

In the months before we left Egypt, my father and I began going to the Nile Hilton more frequently, almost every day. The routine was always the same: A leisurely taxi ride from Malaka Nazli to the Corniche, then a stroll along the Nile. My Cat Scratch Fever was better, and I enjoyed the short walk in the sun with my dad.

Docked a few feet below us, we could see hundreds of small boats, mostly fishing vessels but some pleasure craft. Every once in a while, my father would pause; it was because his leg hurt, but he didn't seem to mind. He was so in love with the Nile, its vast, hypnotic calm, the boats that we could almost touch.

Pointing to a vessel, he asked me wistfully if I wanted to go for a boat ride.

But it seemed so little and slight and unsteady bobbing on the water that the thought of climbing aboard made me shiver. I shook my head no, gripped his hand even more tightly, and pulled him along, away from the boats and the Corniche. He chuckled, and resumed walking. I much preferred the large, imposing feluccas that we glimpsed in the distance, their gracefully curbed sails billowing in the breeze as they drifted along the water.

I was relieved when we finally made it to the safety of the Hilton. We entered through a discreet side door, bypassing the bustling sun-drenched lobby with its glass entrance and its loud tourists and its sub-dued diplomats who could be found day or night milling about the hotel. We found ourselves in a dimly lit bar, all dark leather and soft music and cold air.

My father headed toward his favorite booth, the one closest to the bar. I slid in and sat by his side.

He always made the same request of the friendly headwaiter: a Stella, the imported beer. I would help myself to a large bowl of pea-nuts the waiter thoughtfully placed in front of me. They were crisp and salty and delicious, and I felt perfectly content. The wonder of the Dark Bar was that even a child of six could feel at home.

Occasionally my father would let me have a sip of his beer. His hands, which were beginning to be affected by what we'd eventually learn was Parkinson's, shook slightly as he passed me the tall mug. We'd hardly exchange more than a word or two. It was enough merely to sit back and enjoy the gentle piano music and relief from the searing heat outside.

The spell was invariably broken by the arrival of one of Dad's clients. The Hilton bar was my father's makeshift office, where he often held his most important meetings, and there were more and more of them, those final days. The get-togethers with serious-looking men frightened me. I could tell they were startled to see me. They couldn't understand why a little girl would be permitted to join their grown-up conclaves.

The more astute among them tried to befriend me, asking the waiter to concoct some special child's drink. They'd pause, embarrassed when they couldn't remember my name, though my father had taken pains to intro-duce me. I'd immediately pipe up: "Loulou—je m'appelle Loulou."

I didn't say much more than that. I had learned at an early age how to behave around adults. I was with them far more than with children my own age, and it would never have occurred to me to run around or make noise or have a tantrum in the bar. I knew from the time I was a toddler to watch my dad and take my cue from him, which usually meant doing nothing other than smiling and graciously accepting what was offered me—a kind word, a hug, even the occasional gift.

At the end, it was a blue doll. It was from a favorite associate of my dad's, an amiable man I'd met before, who had come to say good-bye and wish Dad well. He handed me a large oblong cardboard box. Both he and my father were smiling as I tore it open, for once shedding my reserve.

Inside was a striking, very un-Egyptian doll with flaming red hair and a short turquoise blue velvet dress. She was tall and thin and not particularly cuddly. Even her short hair had been sprayed and lacquered into a bouffant shape, so that I didn't know exactly what to do with her. She came glued to her own pedestal, and it was hard to hold her and play with her and dress and undress her the way I did my other dolls.

As I busied myself with the blue doll, my father and his friend were intent on their conversation. They ordered beer after beer, so that we lingered far longer than usual at the Hilton. Dad seemed so comfortable in his booth, he looked as if he could happily have stayed there forever.

I tended to view my father's clients as intruders, rivals for the bond I enjoyed with him and the quiet time we shared inside the bar. It was so blissfully peaceful there compared to the panic and mounting chaos on Malaka Nazli.

We had only a couple of months to sell our apartment and liquidate all our possessions. Under the draconian laws of the Nasser regime, we could take nothing with us save for a few Egyptian pounds; my family was allowed the equivalent of $200 for all six of us.

There was no choice but to spend down the money, but there wasn't much we could spend it on. Jewels, antiques, heirlooms, religious icons such as Torah scrolls, even works of art that had been in a household for generations, could not leave the country. The rules were strict and often cruel. Women were forced to abandon even their engagement rings.

Clothes were among the few items deemed permissible to take in large quantities. My family, like many others, embarked on what amounted to a frantic shopping spree. There were endless trips to fabric stores, department stores, and tailors as we sought to exhaust our family savings on suits and sweaters, coats and dresses, Egyptian cottons, sheets, and blankets.

We were traveling to frigid climates. Europe and America may as well have been Alaska or the North Pole, the way we viewed them, places where, if we weren't careful, we could very well freeze to death with our delicate Mediterranean constitutions. "Il neige là-bas tout le temps," my mother remarked; It is always snowing over there.

No one in my family had ever been outside the Levant, not even my father, who saw himself as worldly and seasoned because of his friendships with the British and the French and the Greeks, and virtually all other foreigners who came through Egypt. In fact, after leaving Aleppo as an infant, Dad had never traveled farther than Alexandria.

The snow-encased lands of our imaginings suggested a need to be practical, to opt for wool or cotton or thick flannel. I was taken by the hand from one boutique to another in search of a winter coat. There were none to be had in Cairo, since there was no winter to speak of in this land of perpetual sun, and it was no easy feat, finding a garment to protect me from the frigid days ahead. "Pauvre Loulou," my mother kept sighing.

My dad went to the Mouski, the old textile district, and purchased yards and yards of shimmering brocade. He chose fabrics embossed with silver threads, in backgrounds of royal blue, crimson red, and green. It was partly an investment—he figured there must be a market for these lush, exotic fabrics outside of Egypt. But he had bought such an excess of brocade that tailors were commandeered to make dressing gowns for him and my brothers.

What they produced looked as if it belonged on some Hollywood set out of the 1940s. Leon, who was over six feet tall, found himself with a robe that came all the way down to his ankles, with a matching brocade belt. I only saw him wear it once, the day that he tried it on in Cairo. He stood in front of the mirror gazing pensively at his own reflection. He looked majestic and formidable and, oddly, younger than his sixty-plus years.

The robes and leftover brocade were folded and packed into a large brown leather suitcase, one of twenty-six assorted pieces of luggage purchased for our trip. They were so massive and heavy, they took up practically the entire bag. Once a bag was packed, it was sealed, pad-locked, and moved to another room, while another bag was brought out in its stead. Each bore a tag that read "Famille Lagnado," but with no address, for we had none to give.

At last, at Cicurel, Cairo's leading department store, a helpful saleslady located one flimsy child's woolen coat in the stockroom. It was gray and hopelessly lightweight, with a single button, though it did come with a pretty matching woolen scarf. She assured us it would protect me from the harsh winters of the West. She seemed confident, but her knowledge base was limited to a city where temperatures rarely dipped below 50 or 60 degrees. My new coat was so thin, it folded into a small square and easily fit into a corner of a suitcase that wasn't even entirely mine, since I, alone in my family, didn't have enough posses-sions to warrant my own bag. With that crucial purchase, my mother crisply announced: "Loulou est toute prête"—I was all set.

I settled back in my favorite corner of the living room and surveyed the anxious goings-on.

Nearly all that we owned had to be left behind.

In my dad's case, his passion for white clothes, he knew, would have to give way to the more sober, subdued colors of the world beyond Egypt. Though he was free to take any clothes he wanted, he wasn't going to bring his prize possessions—the white sharkskin suits and jack-ets he had collected over the years. Like Mom's and Suzette's collection of white shoes, they suddenly seemed superfluous, a relic of a life that was ending. His secret stash of red tarboosh was also left behind.

Because most clothes were handmade, several times a day my mom and sister were off to fittings at the dressmaker. They would return car-rying enormous packages. Dresses with full skirts were all the rage, and Suzette was having them made in what seemed like every fabric and color. I'd gaze enviously at a striking cherry-red corduroy dress, with an enormous flare skirt, wishing I could have one exactly like it. My mother, who had always been more austere, also came home with extravagant new clothes that weren't at all what I knew her to wear. She modeled her new royal blue polka dot dress, and I stared, discomfited, wondering why my staid mother was suddenly wearing polka dots.

Would twenty-six suitcases even be enough? Sometimes one of my parents or siblings would hand me an item they'd helpfully picked out for me. It was a sweater several sizes too large, or thick pairs of woolen slacks, or more flannel pajamas and cotton undershirts. What I was getting seemed nothing like the fineries everyone else was purchasing.

No one seemed especially concerned as to how lost I was feeling. They had adult worries, and adult purchases to make, and adult goodbyes to extend, and my six-year-old's angst seemed too trivial to bother with.

The only creature who still seemed attuned to my needs was Pouspous, who was once again my constant companion. Any fears about letting me stay close to the cat had been set aside, or perhaps my family was simply too preoccupied to keep us apart.

She seemed to delight in the very mayhem that left me so anxious. Pouspous darted in and out of open suitcases, nuzzled her nose into the brocade robes, found in the new luggage a million nooks and crannies in which to hide, and generally tended to amuse herself in spite of—or perhaps because of—the tumultuous goings-on.

Pouspous had been with us since I was born. Like all the other cats that preceded her at Malaka Nazli, she was a stray who had wandered from the alley into our ground-floor apartment and made herself at home. She immediately gravitated to Leon—all cats did. He taught her to appreciate "human" food—since there was no cat food in Cairo, no cans of Little Friskies or Purina Cat Chow in a country where the average person could barely afford to buy a loaf of bread. Whatever Dad ate, Pouspous ate, which meant that when he was enjoying his favorite snack of cheese and Egyptian peasant bread, the cat would nibble on small cubes of cheese and bits of bread.

I almost expected the cat to lap up the hot tea Leon drank at each meal in a tall glass, but instead, he would pour fresh milk for her into a small porcelain saucer and place it near his tea. When she was with my father, Pouspous never ate on the floor.

As our departure date approached, my mother grew more nervous. I saw her one morning stuffing her wedding gown, veil and all, with its yards and yards of satin, into one of the suitcases. It was impossible to fold, so she laid it out almost full-length in the enormous leather bag. To protect it, she covered it with an old fur coat I'd never even seen her wear.

"Ah, c'est de l'astrakhan," she told me proudly, as she noticed me peering at it. Like the wedding dress, it was from another era, when Persian lamb was the height of fashion and luxury. My father had given it to her as a gift during one happy interlude in their troubled twenty-year entente.

Finally, I saw her take a round, dark gray steel box, peer at the contents, and place it delicately under the wedding dress. Whatever was inside the box would be protected by multiple layers—fur, satin, lace, and steel.

My father took over a couple of suitcases and stuffed them with his favorite items. One bag, for example, was crammed with his books—prayer books, dozens and dozens of them, some so old and tattered they could have been holy relics. Their pages were so frail and withered that I didn't dare go near them, for fear that they would tear at the slightest touch. I knew from an early age that they were my father's most precious possessions, so there was never a question of leaving them behind—not even the oldest, most battered one among them.

Another was devoted to canned goods, as if the cities where we were venturing could lack edible food, and the family would go hungry. My father, who did business with a number of canning factories, took it upon himself to collect the staples he felt would help us survive this journey.

After our morning stop at the Nile Hilton, we'd take taxis to distant warehouses on the outskirts of the city. Dad would leave me in the middle of a factory to meet with the owners behind closed doors. He emerged with cases of canned mangoes, guavas, peaches, and pineapple. As we stocked up, my father had a sense that, even in the worst of times, without a home and bereft of money, the family at least wouldn't starve.

Most important were the sardine cans we were collecting to take with us, my father's passion.

One morning, my dad made it a point to leave the house earlier than usual; we skipped the Nile Hilton. Instead, the cab took us directly to a factory located on the outskirts of the city. He had with him a small pouch I had seen him place in his inside jacket pocket in the morning. He had shown my mother its contents before leaving the house.

At the factory, Leon made his way to the manager's office. This time, he had me come inside, though I was instructed not to say a word. After shutting the door, the two sat down together at the desk, and my father removed the pouch from his pocket, and emptied its contents on the table. The manager's eyes widened; what came tumbling out were a half dozen small gold ingots and one sapphire ring.

It was my mother's favorite, and it dated back to the early days of the marriage. Dad had surprised her by offering her this elegant and distinctive ring, which she wore in addition to her wedding band. Gold and studded with diamonds, its centerpiece was an enormous azure stone as blue as the Mediterranean, or the shade of Baby Alexandra's eyes.

My mother was heartbroken at having to part with it. My father had persuaded her that only by giving it up was there any hope of saving it.

My father and the factory owner exchanged a few words, sotto voce. I couldn't make out what they were saying, other than that Leon seemed more anxious than usual, and the man was trying to reassure him. Finally, the two rose, and my father motioned to me that it was time to leave. To my surprise, the owner scooped up the gold and the ring, put them back in their pouch, and promised Dad he would take care of the matter. As we left, he had us walk through the factory and pick out cans of our favorite preserves—orange, pear, guava, apricot, strawberry, fig, even rose petal—to take with us.

When we returned some days later, the manager was expecting us. He smiled broadly while pointing to half a dozen cans on his desk. My mother's ring had been hidden inside a tin of marmalade and sealed within the factory itself so that it was indistinguishable from any you would find at a grocer's. The gold ingots had been similarly stashed inside different cans of preserves. It seemed a perfect way for us to smuggle at least some of our wealth out of the country. My father scooped up the tins. We hailed a taxi and returned to Malaka Nazli, where he told my mom and siblings what he had done.

These were terrifying days, when my family worried that, any moment, authorities would be knocking at our door, prepared to stop us from leaving on some pretext or another, or worse still, ready to haul one or more of us to prison.

That is why my father agonized about his ploy. He would eye the cans containing the gold pieces and the ring, wonder if his scheme was as foolproof as he thought, and worry that there was terrible risk involved.

At last the twenty-six suitcases and duffel bags were sealed shut and loaded onto a van that was to precede us to Alexandria, the first leg of our journey.

On the eve of our departure, my father took the cans containing the gold and Mom's ring and told us quietly he couldn't go through with it. It was simply too risky, and the possibility the authorities would somehow find the ingots and jewelry posed an unacceptable danger to us all.

We had heard terrifying stories about the customs inspectors, and how thoroughly and ruthlessly they searched everyone who left, especially Jews. One woman, a seamstress, had hidden her engagement ring in a small iron she used to press hems. Then, in a touch of ingenuity, she had taken gold coins, covered them with cloth so that they looked like ordinary buttons, and affixed them to a dress. As she prepared to leave, an inspector had examined the iron and found the hidden ring. Minutes later, he had stripped the phony buttons off the dress. He proceeded to tear through nearly every item of clothing, even ripping the shoulder pads off her husband's fine, hand-tailored suits, in search of hidden jewels. Miraculously, she had still been allowed to leave, albeit with none of her valuables.

At the table, Dad opened a can of orange marmalade. There, precisely as the factory owner had vowed, was the sapphire ring. He opened another can and found two gold nuggets. The other cans yielded the rest of the cache. My mother took her ring and washed it under the sink, then dried it lovingly with a towel.

Leon told my mother he would have to take the ring back and dispose of it. Nothing was said of the ingots, but like the ring, they vanished, never to be seen again, and none of us were ever told exactly what our dad had done with the beautiful gold pieces.

When it was time to leave Malaka Nazli Street, Pouspous was all I could think about. My father held the cat as I said good-bye. "Can't we take her with us?" I asked one member of my family after the other. No one seemed willing to level with me, to tell me plainly that we were leaving Pouspous and Malaka Nazli forever. But I knew, of course.

On the dining room table were odds and ends we had left till the last minute, uncertain what to do with them. In one corner was my leather *cartable,* my first schoolbag, which I'd carried so joyously to my classes at the Lycée de Bab-el-Louk, feeling so grown-up.

My father came to lead me out of the house. We always walked hand in hand; because of his limp, he walked so slowly that it was easy to keep pace, the way I couldn't with other adults. I went to stroke Pouspous one last time.

"She will be fine," he reassured me, in that mild tone he used only for the most important subjects.

My father could be gruff and imperious on minor issues, but on all that really mattered, he was astonishingly gentle.

Pouspous preferred staying put, he told me. "Elle veut rester ici, elle aime le Caire"—She wants to stay here, she loves Cairo—he said again and again. He tried to soothe me by painting a picture of the cat lingering behind in our deserted apartment, a Cairene at heart, unwilling to give up her home. She would have every room to herself. She would be able to sun herself on the balcony, sleep in any corner of the house she pleased, eat the mountains of food we were leaving behind, and be the queen of her domain.

Pouspous had no desire to leave, no desire whatsoever, my father kept insisting.

But I couldn't stop crying. At last, my father said that he would have a talk with the cat and persuade her to come join us in Alexandria. She needed a few more days to get ready, that was all, and then she would meet us at the port. He offered me the large white handkerchief he always carried in his pocket to dry my tears. I decided to take him at his word.

Pouspous didn't even look up as we left. She had retreated to her favorite spot on the balcony, and sat curled up in her favorite position. As we all trudged out, she continued placidly watching the street life of Malaka Nazli.

IN ALEXANDRIA, WE HEADED for a small hotel where we had some-times stayed over summer holidays. It was a simple guesthouse, close to the sea, with small rooms that felt even more cramped with our piles of luggage. I tended to associate Alexandria with carefree vacations and fun-filled days at the sea, so I felt confused. No one in my family spoke of going to the beach. No one said much at all.

I held my blue doll at all times, though I wasn't used to her hard edges, and didn't find her as huggable as Pouspous had been.

Suitcases and duffel bags took up the better part of one or two of the bedrooms. In addition to the letter tags that hung from the handle, my brothers had painted FAMILLE LAGNADO in white block letters across the top of each suitcase to avoid the possibility that some other family on the run would end up with my mom's old wedding gown, or my father's yellowing edition of the Talmud.

On our last night in Egypt, my father took me for a walk. Hand in hand, we made our way across the boardwalk, passing one beach after another. Every once in a while, he would stop and turn and face the sea. He didn't say a word. There were countless cafés, and even in March, we could see nighttime revelers relishing the breeze of an Alexandria evening, smoking and drinking beer or arak, the liqueur whose smell I loved, but which was so strong I couldn't touch a drop. My father set-tled on the very last café.

It was almost deserted, and to my surprise, we sat indoors, which was dimly lit and quiet, almost like the Dark Bar. My father motioned toward the waiter and ordered a beer. The waiter, who seemed to know Dad well, returned with a tall mug of cold beer and a special treat for me—a tall dark red drink. It was a strawberry soda, and it was on the house, he announced as he plunked the glass in front of me. "C'est pour la petite," he told my father amiably. His cheer contrasted with our bleak mood.

The deliciously sweet, syrupy drink almost made up for the sadness of the night, and I felt comforted as I took small sips. We resumed our walk on the Corniche and returned to the hotel. Dad kept looking at the sea, and he never let go of my hand.

In the morning we made our way to the dock, where our ship was expected to sail around noon. The waiting area was surprisingly crowded. Though it was a hot day, my mother had bundled me up in several layers of clothing—a dress, two flannel undershirts, a sweater, and then another sweater.

I had been coached to be on my best behavior with the inspectors, to smile and introduce myself.

When my turn came, I saw a tall man motion to me with his finger to approach. "Loulou—moi je m'appelle Loulou," I told the man in uniform, trying not to sound frightened. I felt lost under my pile of clothing. I wasn't even sure he could see me clearly. The man smiled and waved me through, sweaters and all.

There were many other families like us, sitting in small chairs, surrounded by mountains of suitcases. They spoke a dozen different languages, Arabic and French, of course, the two most common languages, but also English, Greek, Italian, and Spanish. But that was Egypt, of course. Or it had been. Suddenly, "foreigners" weren't welcome in the very place where most of them had felt so profoundly at home.

A woman seated across from us was carrying a small portable cage. I saw that it contained a cat. She would occasionally open the cage door and stroke the meowing cat and mumble a few words to persuade it to calm down.

But where was Pouspous?

I kept looking around to see if she were coming, as my father had promised. I was distraught. "We could have taken her with us like that woman," I told my family reproachfully, and started to cry all over again. My mother ruefully agreed that putting Pouspous in a cage and carting her out of Egypt would have been a good idea.

"Elle ne voulait pas laisser le Caire," my father told me; She didn't want to abandon Cairo. He had adopted his mild tone of voice again. He made it sound as if he'd had a rational conversation with my cat, and she had clearly conveyed her wishes to stay exactly where she was.

As we boarded the boat, an inspector made us sign one last official document.

It was known as "un Aller sans Retour"—we were promising to leave and never come back.

BOOK II

THE EXILE

• • • • •

PARIS, AND THEN NEW YORK

1963–1982

CHAPTER 13

The Jewel Within

We had barely drifted out of Alexandria's harbor when I heard my father cry, "Ragaouna Masr"—Take us back to Cairo.

It became his personal refrain, his anthem aboard the old converted cargo ship that rocked so violently as it crossed the Mediterranean that we couldn't bear to stay even for a moment in our inexpensive lower berths, but would slip upstairs to the relative comfort of the upper deck. There, seated on high-backed canvas chairs aligned in straight rows, we'd spend our days and nights, unable to sleep or eat or do much more than try to look back at what we had left behind, or ahead to what awaited us.

Both suddenly seemed blurry.

Past and future looked as vague and out of focus as the lone photograph that survives of my father and me aboard the *Massalia*. There we are, huddled on the upper deck, while behind us, dozens of people sit silently watching the sea. It is like a scene from a cruise ship ad gone awry: none of the passengers seem happy, least of all my father. In his dark felt hat, jacket, and tie, he is dressed far too formally for a sea

Leon and Loulou on the Massalia.

crossing. He stares straight ahead at the camera, looking sullen and worn and, for the first time, old. I share his melancholy. My head is lowered, my eyes are downcast, and if it is possible for a six-year-old girl to feel defeated, then I look as if I, too, have lost my purpose. Perched against his shoulder, I am holding on to him, in need of his protection even now that he may be incapable of giving it.

It was my brother who snapped the smudgy image in March 1963, using a cheap portable camera.

My father talked obsessively of Malaka Nazli, as if expecting me to grasp all we had lost. After months of frenzied activity, there was nothing left to do. Nothing except to sit back on our deck chairs, and gauge how it had all come to this—decades in the life of a family, reduced to two dank cabins situated too close to the roaring engines of a small unsteady boat, along with the twenty-six suitcases that contained all their worldly belongings.

"Ragaouna Masr," my father kept shouting. He had lost all inhibitions, and for a man whose life had exemplified elegance and propriety, any sense of decorum seemed gone. He would cry out when he sat alone, and he'd cry in front of other passengers, when we were with him and when we weren't, inside the privacy of our lower berths and out in the open air. Oddly, no one seemed to mind—or even to find it strange—the sight of this irate old man, at times yelling, at times softly moaning, Take us back to Cairo. He was only saying what they felt in their hearts.

It was at that moment, as the *Massalia* bobbed up and down on waters as green as my father's eyes, that I realized how different he looked. I had first noticed the change, a resignation in the way that he drank his beer, ever so slowly, at the café in Alexandria. The jauntiness and self-confidence that were such an intrinsic part of his persona were gone.

It was all so disorienting, as if the man I called my father was an impostor—a desperate stranger I had never met before.

My father wasn't fifty-five, as his exit papers said; he was sixty-two or sixty-three and looked and felt even older as our ship set sail. He wondered, as he experienced the rush of pain that came and went, came and went, in his hip and thigh, how he was going to begin anew, find work, and support a wife and four children, including a little girl who clung to him for dear life.

As the *Massalia* chugged along the Mediterranean, stopping first in Greece, then Italy, I had virtually no dealings with Suzette. My sister kept to herself, unwilling to wallow in the collective despair. Unlike my father, she felt nothing but relief as the boat edged out of Alexandria. Egypt had been a nightmare, even before her arrest and the furor over her conduct. The country that my father already missed so much, that he cried to be allowed to enter again, didn't exist anymore as far as my sister was concerned, and hadn't for years.

Free at last, she relished the idea of a future in the West without the restrictions and stifling mores of the Levant.

Meanwhile, my oldest brother, César, constantly seasick, found he couldn't think clearly. He didn't share my sister's boundless optimism— or my father's boundless despair—but remained suspended somewhere

César on the deck of
the Massalia.

between the two. He kept running upstairs for fresh air, unable to stand the claustrophobia of our lower cabins and the constant thud-thud-thud-thud of the engines, and the waves crashing against the porthole. I never seemed to see Isaac or my mom.

Late one evening, we drifted into Genoa, the last port of call before France. There at the docks, waiting for us, was Salomone, the glamorous cousin from Milan I had heard so much about. As tall as my father, elegant and exquisitely dressed, he was an imposing figure. Everyone in my family—my mother in particular—seemed thrilled to see him.

It was their first reunion since Salomone had left Egypt fourteen years earlier, in 1949. Married and a father of three, he was presiding over a growing import-export concern, and his business stretched from Europe to Africa. Yet his home base was still Milan, the city of his youth and memories. Ghosts of his parents and little sister lurked in every corner. Salomone lived in an apartment close to the Duomo, where the family had lived shortly after he was born, and not far from the square where Mussolini had founded the Fascist Party. This was the school Violetta had attended and where she'd written her first poems, and over here was the prison where she and his parents had been held. And there, of course, the station where they had boarded the train to Auschwitz at dawn, never to return.

We followed Salomone to a nearby café, popular with sailors. The only bit of cheer within the stark, poorly lit interior was the overabundance of Easter decorations. All around the eatery, I could see displays of Easter eggs. They hung alluringly from ceilings or were stacked side by side along high shelves, dazzling in their gold and silver foil casings, and artfully tied with large satin ribbons. The eggs were made entirely of chocolate, and several were at least a foot tall.

I found myself obsessing over what lay beneath all the layers, and left the table to inspect them more closely. My parents and cousin were so deep in conversation, they seemed oblivious to my comings and goings: they were busy reminiscing about their years together at Malaka Nazli, Salomone and my father, along with my grandmother Zarifa, and at the end, the lovely young stranger who joined their household, my mother, Edith.

But they also flashed forward to our family's current plight and what we were going to do now. My cousin had tried to arrange for Suzette to live with him and his family. He had gone to see high-ranking government officials to obtain the proper papers, to no avail. The authorities wouldn't allow my nineteen-year-old sister to remain in Italy. As far as the rest of our family, while Milan was appealing because of our bond with and love for Salomone, it simply wasn't an option. There were many Egyptian Jews who had settled in Italy, but most were able to claim Italian citizenship, however tenuous. They wangled their way into Rome or Milan by stating they were "half-Italian," brandishing a wife's Italian passport, or a parent's or grandparent's. We were stateless, which meant our movements were severely limited.

We only had permission to go to France, where we would be allowed to stay a few months until we found a permanent refuge.

As I continued my tour of the café, eyeing the chocolate eggs, I looked up to see Salomone towering over me. He was, like my father, a man of few words. He asked me which one I wanted. The café, which had felt melancholy and dim, suddenly seemed flooded with light. I wasn't sure what to do. I knew that I couldn't point to the largest, most extravagant egg, but I didn't think I had to settle for the smallest egg either. I stood there, incapable of making a decision.

Salomone finally extended his long arm to the ceiling, plucked an egg from a top shelf, and handed it to me. It was elegant and massive, wrapped in silver.

"Ça va?" he asked me.

I nodded, dazed by his gesture, and returned to the table, brandishing my Easter egg like a trophy. When it was time to return to the boat, everyone rose, my father and my cousin embraced. Salomone lingered briefly by my mother, hugged her tenderly, waved to us, and climbed into his car and drove away.

As we walked, I began to unwrap the egg. I had to peel off layers of foil, tissue paper, and bits of ribbon, until at last I saw the outline of an immense globe-shaped milk chocolate. I broke off a piece. The egg was hollow inside, and I realized there was a gift within the gift—that deep inside the Easter egg, a prize had been stashed away. My hands finally retrieved a cellophane pouch containing a pair of golden earrings, small and simple and beautiful.

"Tu crois que c'est de l'or?" I asked my father; Do you think it is real gold? My brothers burst out laughing, but my father wouldn't say yes or no. I stuffed the earrings in my pocket and continued walking. We could hear the Greek crew amiably shouting to everyone to hurry up and come aboard, lending passengers like my dad a hand walking the rickety gangplank.

At last, the boat floated into Marseilles. Exhausted from the voyage, we had no means to check into a hotel but hurried, our twenty-six suitcases in tow, to catch an overnight train to Paris. César left us to explore the station. He wore his prized black leather jacket, a *blouson noir,* one of his last purchases from Egypt. It was to be his passport to the stylish world of the French.

As he wandered aimlessly, he suddenly found himself flanked by two plainclothes officers. They pushed him against a wall and began to frisk him. They had noticed him roaming the station, dressed all in black, and had mistaken him for a "Blouson Noir," one of the North African gang members who were terrorizing France and were involved in protest actions against the unpopular war in Algeria. Any young man who fit a certain physical and ethnic profile fell under immediate suspicion, and my oldest brother, with his classic Middle Eastern good looks—dark curly hair, brown eyes, fair skin, and black leather garb— could easily pass for an Algerian immigrant.

Pointing to an unmarked car, the officers asked him to accompany them for questioning. His eyes widening, César shook his head no, no. He was certain that if he obeyed the men and followed them into the car, he would never see us again. Trying to gather his wits, my brother explained that he was indeed a refugee from North Africa—not Algeria but Egypt. He urged the officers to locate our parents, who were in another part of the station. "Mon père est là-bas, avec ma mère et ma famille," he pleaded; My dad is over there, with my mom and the rest of my family. But the officers seemed uninterested in finding any of us, and César had no choice but to keep trying to talk his way out of his nightmare.

He didn't belong to any gang, my sixteen-year-old brother assured them again and again. His outfit, his leather *blouson*, was simply a nice jacket he had picked up in Cairo. They still cast a cold eye on his professions of innocence. After peppering him with dozens more questions, they reluctantly let him go, and sped away in their unmarked car.

My brother, still shivering under his *blouson noir*, joined us as we boarded the train to Paris. He didn't breathe a word about what had happened.

He had been in France exactly one hour.

I fell asleep by my father's side, clutching what was left of my Italian chocolate egg. Sometime in the middle of the night, somewhere in the middle of France, the train came to a sudden halt. The rail workers were on strike, we were told. We were caught in one of the country's legendary union actions. There was no choice but to remain in the darkened locomotive.

The Marseilles-to-Paris journey had turned into a frightening web of deserted open rail yards, long dreary waits, and trains that went nowhere. We were exhausted and cold, and we could only wonder at our first taste of life outside Cairo.

AT LAST, THE ENDLESS night journey across France came to an end. We were in Paris and it was morning and there was light.

From the station, my father telephoned a contact at the relief agency helping the flow of Jewish refugees from the Arab countries. We were listed as "stateless" on all our travel documents, and we didn't know our final destination. Dad anxiously inquired what was in store for us now that we had left Egypt. He had an exhausted family on his hands, and he wasn't well himself. There was also a six-year-old child, "une petite qui est très fragile," he informed the agency official, his voice nearly breaking from tension and fatigue. We were told to report to our temporary lodgings in the tenth arrondissement, at the Violet Hotel.

Leon's identification papers, Paris, 1963. This nationality was "a déterminer"—to be determined.

I liked the sound of it. I expected a building of lavender walls and lilac floors rising beneath a mauve-tinted sky. Instead, we found ourselves staring at a dingy and singularly charmless establishment that was all gray and discolored broken bricks and stone. Our rooms were situated on an alleyway known as the passage Violet, a narrow lane of fabric stores, fur workshops, and small factories that made buttons and dolls. I vainly scanned the little street for a speck of purple, but there was none that I could see.

It was even worse inside.

Home was now a couple of rooms containing six beds and the twenty-six assorted suitcases that had followed us from the station. Because of their bulk and size, they turned us into virtual prisoners of our hotel, taking up so much space we could barely walk without stubbing our toes or bumping into one another.

Our rooms were on the second floor of an annex that was, if possible, even more decrepit than the main hotel building. We had to climb a rickety flight of stairs, a painful and awkward undertaking for my father.

It was in Paris that Papa's cane, packed as an afterthought when we left, made a surprise reappearance. I had rarely seen him use it in Cairo, where years in the care of top doctors had enabled my father to attain a fair degree of mobility and independence. But he now needed it to get upstairs, and then to get down again.

We didn't bother to open any luggage. It wasn't clear if it was because we were too depressed, or because it would have been pointless. The bags that my older sister had packed with so much excitement, cramming them with her new wardrobe, were now a source of exasperation. Why can't we unpack? she kept asking my father. Why do we have to keep the suitcases locked as if we were about to flee again?

He shrugged as if to say that at nineteen going on twenty, she was old enough to figure it out. Paris was only a stop on a long, as-yet-unfinished journey. When we arrived in France, at the end of March 1963, we were still in the same limbo status indicated on our luggage tags from Cairo, "Famille Lagnado," but no discernible address.

Around the corner was the rue du Faubourg Poissonière, a narrow, windy street that looked exactly like every other narrow, windy Parisian street, with one major difference. Poissonière and the area around it had catered to generations of Jewish refugees who had fled any number of countries that no longer wanted them.

The cultural and historical landscape had changed over the years, but the story of exile and persecution was numbingly the same.

France had historically been a transit point for refugees, a role it was re-creating assiduously now with the flow of Jewish families like my own. In the 1930s and '40s, Jews fleeing the Nazis converged on this small strip near our hotel, some opening ateliers and plying the fur trade. Later, in the 1950s and 1960s, Jews seeking to escape the violence and turmoil in the Middle East unleashed by the creation of Israel descended on the faubourg. The area attracted immigrants from Algeria, Tunisia, Morocco, and Libya who were housed in the shabby residential hotels that did a booming business. Joining them were another class of refugees, once-prosperous families such as mine, from Cairo and Alexandria, who went overnight from riches to rags, and were in a state of shock at the seaminess of their new lives.

It was possible to walk around the rue du Faubourg Poissonière and hear a cacophony of languages—old furriers who still spoke German and Polish and Yiddish, refugees from the Maghreb who felt comfortable conversing in their native Arabic. That left French to the streetwalkers and transvestites who plied their trade not far away, by the boulevard St. Denis.

Paris had a relatively efficient, coordinated system of social service and relief agencies dedicated to helping refugees like my family. Funded by private philanthropists such as the Rothschilds, as well as deep-pocketed American Jewish organizations, the French groups tried to lessen the trauma. Refugees were immediately given a free place to live—typically a room or two in an inexpensive hotel—along with subsidized meals. They were put in contact with officials who would help find them a permanent home somewhere in the world.

The main agency helping us in Paris was the Cojasor, an organization that had once aided Holocaust victims. HIAS—the Hebrew Immigrant Aid Society—was the other major group in charge of our fate. Headquartered in New York, but with satellite offices around Europe, HIAS's mission was to help repatriate Jews forced to flee because of the tumult in the Arab world.

From our first day in Paris, we trooped from one to the other, one to the other. Cojasor, which was helping us navigate life in France, assigned us our own social worker, kindly Madame Dana, to help us through the rough spots. HIAS was trying to look ahead and help us decide where we wanted to settle for good.

The choices were clear: either Israel or America. Those early weeks, we leaned strongly toward Israel. My sister dreamed of finding Nonna Alexandra again, while for my father, there was the hope of joining his relatives, including his ailing brother Shalom and Marie, his little sister.

Dad went regularly to the Cojasor to collect our allowance, which amounted to eighteen francs daily, or three francs a day for each of us. That austere budget was supposed to cover all our needs, from food and clothing to the occasional movie or treat. The money was handed directly to my father, who was tasked with deciding how to spread it among the six of us.

We had arrived in France with exactly $212—the sum total that we'd been allowed to take with us out of Cairo—and much of it was gone.

For Dad, who had spent a lifetime investing in the stock market and building a nest egg, what was most painful was finding himself destitute, dependent on charity for himself and his family to survive. My father was used to giving alms, not taking them.

He had dreaded that moment but had been helpless to avoid it. We heard, of course, of families who had been able to smuggle their fortune out of Cairo by using trusted intermediaries or couriers to transport their cash and jewels, or by establishing secret Swiss bank accounts. But most of the Jewish refugees of the Levant found themselves in exactly the same straits: they saw their social status and wealth vanish overnight. They went from being solid members of the bourgeoisie to beggars.

Now, Dad had to mediate between the demands of my siblings, who wanted pocket money to enjoy what they could of Paris, and the family's basic needs.

He spent almost nothing on himself. The boulevardier of Cairo now wandered around in a faded raincoat, which became increasingly battered. In spite of the shopping sprees in the months before we left, Dad had never thought practically about what he would really need in the world beyond Egypt. It almost never rained in Cairo, so raincoats were a rarity, and even an ordinary umbrella was an exotic object, almost impossible to find even at Cicurel or the other great department stores.

He asked the Cojasor for help in purchasing a new raincoat. The agency refused, though his request was passed along in global tele-grams to HIAS offices overseas. The answer was always no. Rebuffed and humiliated, he stayed inside our hotel room. He spoke up only to order my brothers to pray with all the authority he could still muster. He seemed anxious that they maintain the old rituals, while they seemed less and less interested in doing so.

My sister, who had argued so vehemently for us to leave Cairo, was now complaining the loudest about France. Like Dad, she found our *nouveau pauvre* status almost impossible to bear. The Paris we were inhabiting had nothing to do with the Paris of her dreams and literary sensibility. Suzette had never imagined being penniless in a city whose boundless charms required sizable sums of money to enjoy them.

She was in a perpetual funk about our reduced circumstances—the cramped hotel room where we were constantly on each other's nerves; the drab neighborhood that didn't interest her in the least; her inability to work or attend school because she didn't have proper papers, and besides, we could be on the move again any day, so what was the point of going to university or getting a job?

Our identity was reduced to a number, Dossier #45,135 of the Coja-sor, filled with case notes by Madame Dana and one dismal word, *state-less*.

Unable to stand our new digs, Suzette left the hotel early and wan-dered around Paris by foot; there was no money for the metro or a bus. She couldn't shake the despair that had overtaken her, the sense that she was somehow responsible for getting us into this mess: if it weren't for her arrest, we would still be in our sun-drenched homeland, living in a real house, with friends, and furniture, and money.

A month after our arrival, my mother received a letter that would plunge us even deeper into darkness.

Nonna Alexandra, of the hard luck and the tender soul, had died.

The flower on the mountain—our edelweiss—had passed away weeks earlier, while we were still in Egypt. The very morning we heard about Nonna, Dad had gone to Cojasor to say we wanted to move to Israel, where we hoped to be reunited with family. Later that afternoon, he returned and told them apologetically that we had changed our minds. We didn't know anymore where we wanted to go, he said truthfully.

The devastating news had arrived in an airmail letter from Oncle Félix, who was now working in Geneva. He hadn't told us what happened for some weeks, he conceded, to spare Mom. After all, there was "nothing left to be done," and it seemed pointless to risk interrupting our plans to emigrate. He, of course, had been in Switzerland, hundreds of miles away from my grandmother, even as her health and hope were failing, and he hadn't seen her in some years. It had taken his estranged wife, Aimee, to inform him that his mother had been rushed to the hospital.

He caught a plane home only to arrive—as always—too late.

Alexandra of Alexandria, the old woman who was more like a child, the grandmother who needed such intense mothering of her own, had died alone and bewildered in an institution, as solitary and lonely a figure in her narrow sickbed as on those walks, six years earlier, among the orange groves of Ganeh Tikvah. "She is finally at peace, after an entire lifetime that was for her nothing but a lifetime of sacrifice and suffering and misery," Félix said in his flowery two-page letter. None of this was especially comforting to my mom. My uncle didn't bother to say precisely when Alexandra had died, what day, what month, from what, whether she had been ill a long time or had suffered a sudden decline. But there was one line in Félix's letter that seemed genuinely aimed at providing comfort:

"I have been told that she was asking about you and your children until her last moment on this earth."

The news rendered my mom almost mute with despair. Trapped in Cairo and now Paris with the five of us, separated from the one person whom she had loved utterly and completely, she hadn't even been with Alexandra when she died. She hadn't seen the body. She had missed the funeral.

She would never see or speak to Nonna, or listen to her sing some cherished, long-forgotten Italian love song, she wouldn't hold her fragile form or comb her silky hair, turned white from sorrow; and now, with neither the means nor the ability to travel, she wouldn't even be able to pay her final respects at her grave.

It was almost worse than when her other Alexandra had died.

Madame Dana at the Cojasor was the first to notice the change. The social worker noted in her case files how broken down my mother seemed, how unkempt and neglected: it was clear she had lost interest in her appearance. Madame Dana was worried about my mom's passivity, the fact that she seemed to agree with whatever was told to her. She had no will of her own anymore—no will to say yes, no will to say no, no will to demand, no will to object. What could a social worker do to help this soft-spoken, intensely sweet woman, old before her time, toothless before her years, who had once been beautiful but was no longer so?

Suzette now found Paris truly unbearable. She had lived for the moment when she would leave for Israel and find Alexandra. But what was Israel without Alexandra? A promised land bereft of any promise. In a world without my grandmother, every country lost its appeal, and there was nowhere Suzette could point to and say, Yes, this is where I want to live.

Even my father, who had resented the influence Alexandra wielded on Mom and the rest of us, found himself shaken by all the deaths—his brother Raphael, his sister Rebekah, and now his ill-fated, troubled mother-in-law. Of all of us, Dad had wanted to go to Israel the most, yet he too was now wavering.

Her death plunged us into a kind of pathological indecision. We tried to stall both the Cojasor and HIAS, as we struggled to come to a decision. We went ahead with an application to settle in America, though it was by no means clear that was where we wanted to go. Discussions among my siblings tended to be intensely emotional or numbingly repetitive. I heard the same hopes and misgivings voiced again and again. Tensions would flare, there'd be yelling, and then calm would again descend, and a consensus seemed close.

Israel would have been a simpler choice than the United States, which entailed an arduous application process that could take months. It was also where most of our extended family had settled. Though Nonna Alexandra was dead, we had dozens of other relatives on my father's side ready to welcome us. It seemed the ideal place to begin again and rebuild our lives, or so one argument went.

On the other hand, my brothers feared the mandatory military service, a near certainty if we opted for Israel, a country under constant threat of war. Plus, work was said to be tough to find, the pay meager, the possibilities for advancement limited.

To say nothing of the scorpions. We risked coming face-to-face with them, since we were bound to live in tents. I shuddered merely thinking about them.

Hope lay in America, César proclaimed, espousing the vision of the carefree, fun-loving culture he had glimpsed in Cairo movie theaters. My brother had watched, mesmerized, a clip of Chubby Checkers performing "The Twist," and it was akin to a religious conversion. America beckoned, not because of its promise as the proverbial land of opportunity but because of its allure as the land of Chubby and Elvis, where everyone danced the Twist and swayed to rock and roll. In the spring of 1963, America seemed to hold so much of what we craved—peace, a home of our own, security, and the possibility of amassing wealth again and recapturing our lost life.

I kept hearing references to streets paved with gold. Naturally, I took them literally. I closed my eyes and tried to picture cars gliding over a shimmering New York City roadway, while I strolled along a golden sidewalk. What a lovely change from the slate gray streets of Paris.

My parents seemed helpless to stop the bickering. Neither wanted to risk losing a son to the draft, which was the defining drawback to Israel. America seemed to pose few such safety concerns. There was no talk, or none that reached our ears, of the growing conflagration in Vietnam. We had no inkling that within a year or two, there would also be a draft in America.

With the exception of weighing in on the scorpions, my mother was quiet. Still mourning the death of Alexandra, she had lost her main incentive for going to Israel, or anywhere else for that matter. Despite a husband and four children, she felt rootless and alone.

No one sought my opinion. If they had asked, I would have said without hesitation: Let's go back to Cairo, let's try to find Pouspous.

My father would have to make the final decision, but he, too, was torn. He was also in physical pain. The ache from his leg was at times ferocious. Over the years, my father had continued to correspond with physicians across Europe who had urged him to come see them the instant he arrived. That's why my father carted around Paris the black old-fashioned X-rays he had brought with him from Cairo.

He took them along one day when we reported to the Cojasor. He informed Madame Dana, our social worker, that he was hoping to consult some specialists in Geneva, Milan, and London about his sore leg. Madame Dana looked genuinely startled.

Didn't he realize, she asked, that with our stateless status, we were under orders to remain strictly within the environs of Paris? His only option—Madame Dana shrugged—was to consult doctors here. She was being disingenuous, of course, while we were clearly in denial.

We had no money to see the great specialists of Paris—or Milan, London, or Geneva either.

A free clinic for the poor became our only portal to the world of Western medicine we had dreamed of entering for so long. I, too, needed medical attention: worrisome symptoms of Cat Scratch Fever returned to haunt me shortly after we arrived—or perhaps we hadn't noticed them in those last, chaotic days in Cairo. The strange swelling on my left thigh that Egyptian doctors had blamed on Pouspous seemed especially pronounced, and I was frightened. What happened to the magical cure I had experienced at the hands of Maimonides? Was it vanishing now that we were so far from his holy shrine?

I sensed that my family was also worried. "Loulou est encore malade," my mother sighed; Loulou is sick again. Both my parents wondered how on earth they were going to care for me in a dingy hotel room, in a city where they didn't have a dime and didn't know a soul.

My father's fate and mine became enmeshed once more as we ventured to the free clinic in search of answers—only to find there were none. Dad was getting worse; his steps were becoming more tentative, his movements more restricted. Only a couple of blocks from our hotel lay the majestic open-air boulevards of Montmartre and the Capucines, and beyond them, the even more sumptuous Opera and Haussmann, yet he couldn't enjoy them. Unable to reclaim his role of boulevardier in the city of boulevardiers, he stayed most of the time inside our hotel room and did nothing but pray.

CHAPTER 14

The Missing Birthday

I t was in Paris that my birthday almost disappeared.

We had been there several months, stuck in the Violet Hotel. Fear for my brothers' safety finally clinched the decision: my father had chosen America. But that in no way guaranteed we would be able to go. Our application was winding its way through the mazelike bureaucracy of global refugee resettlement agencies, and we still hadn't unpacked.

My father now occasionally ventured to Montmartre, where an outdoor fair, complete with barkers and games of chance, restored his good humor. His passion for gambling found an improbable outlet amid the small kiosks where for less than a franc and the spin of a wheel of fortune, it was possible to vie for watches, dishes, pocketknives, transistor radios, even jewelry, which I eyed longingly, since all I possessed was the pair of earrings I'd found inside the chocolate Easter egg. The fair became Dad's daily destination, after morning prayers and before joining the sad sacks who gathered at noon at a nearby soup kitchen.

The Cojasor had set up an elaborate system of meals for the waves of immigrants from the Levant. A communal cafeteria on the rue Richer, a block from our hotel, served hot, nutritious meals that adhered to the strictest Jewish dietary laws, all for a nominal fee. Once a day, around noon, elegant socialites from the sixteenth arrondissement descended on a mission to help feed the poor and destitute, which now included my family.

At Le Richer, as we called it, meals were freshly prepared and plentiful—almost too plentiful. Every few minutes, bejeweled volunteers came to our table, offering to ladle more food on our plates from the large tureens they carted up and down the aisle. They seemed more than willing to offer us extra helpings of dishes that tasted foreign and delicious—especially the charcuterie they served once a week with bread and mustard. I'd never tasted cold cuts in Cairo, where the lone kosher purveyor had packed up and moved before my family left.

I found the thin slices of salami and turkey Le Richer offered us intensely exotic, unlike any food I'd ever eaten.

An elderly countess—at least, that is what César said she was—gold bracelets dripping from her arms, precious stones adorning the fingers of each hand, would approach us and always ask the same question, in the high, squeaky aristocratic voice that my oldest brother loved to imitate: "Beaucoup ou un peu?"—A lot or a little?

If we shyly nodded *beaucoup*, she heaped our plates with mountainous portions of meat, rice, vegetables, and potatoes, and if we said "Un peu," she still tried to fill our plates to the brim, so that it really didn't matter how we answered her. We were encouraged to take food home; among the few amenities of the Violet Hotel was a small kitchenette with a *rechaud,* an electric burner where it was possible to recycle leftovers and have them for our supper.

Though the volunteers behaved graciously and made sure no one went hungry, most of us still found Le Richer impossibly bleak. Dad yearned for his simple meals of pita bread and cheese. My sister couldn't get used to the noisy, impersonal cafeteria-style dining. I longed to be seated at our dining room table, sneaking bits and pieces from my plate to Pouspous.

Day after day, we had nothing much to do except report for lunch and rejoin the community of losers who shared our sense of aimlessness.

My reprieve from the world of exiles and lost souls came as a result of my mom's restlessness. After news of Alexandra's passing, she began to take long walks with me, clutching my hand and forcing me to wander with her for hours through Paris. We sprinted down the passage Violet and over to Montmartre and its bustling, low-budget stores, with their foraging and stampeding crowds of shoppers, and continued to walk for miles until the scenery changed, the stores became silent and spare, and the people we glimpsed within them seemed to glide rather than walk.

We crossed boulevards and avenues. We explored alleyways and peered into courtyards. We traversed bridges and islands. I knew only to hold her hand tightly and try to keep up. Her brisk intense pace was so different from my dad's labored shuffle.

The few words she exchanged with my father were to ask for a few coins each day. She would use them sparingly, conscious of how little we had, and treat me to an ice-cream cone or a bag of chips. It had to be one or the other, not both. Each day, I was forced to make what seemed an impossibly difficult decision, as tough in its own way as the choice that had been rankling my parents and siblings, namely, whether to settle in Israel or America.

I, too, couldn't make up my mind, couldn't decide between a cool refreshing vanilla ice cream or the crisp, salty potato chips I'd never tasted before France. I tried desperately to accept the limits placed on my once limitless needs and desires, the needs and desires of a pampered little Levantine girl, used to seeing the world from her charmed balcony facing Malaka Nazli.

Occasionally we walked with a purpose and destination. We ventured to the Prisunic, the French five-and-dime with its bins of bargain fare, or the Tuileries gardens, whose bucolic expanse was so restful. But nothing gave us more pleasure than going to Parc Monceau, the lush, rich children's playground nestled in the seventeenth arrondissement. It seemed miraculous that it was free and accessible to us because so much of Paris was out of bounds—the lovely bistros and cafés, the theater, the opera or the Comédie Française, all cost money we didn't have.

As we prepared to walk through the wrought-iron gates with the gold "PM" insignia, Mom reminded me this was Marcel Proust's playground, and she said it with so much feeling and intensity that I knew I was expected to absorb the magic that she suggested was in the air and the grass and the swings and the sandlots.

I noticed a change in Mom the minute we entered. It was as if we had reached a spot where the oxygen level was higher, and she could breathe again. There, amid the Japanese gardens and rolling brooks, in the shadow of simulacra of an Egyptian pyramid and a Roman ruin where toddlers played before the watchful eyes of their English nannies, my mother seemed to come to life again, and her beauty returned.

As Edith sat on a bench gazing at the elegant young mothers, feeling neither young nor elegant, but distinctly more hopeful, I ran in the grass, fed the ducks in the pond, climbed hills, and watched puppet shows alongside the other children whose outfits probably cost more than all my family now possessed.

The sandlot and the swings were great equalizers. I began to join in their games, strangely at ease with the privileged girls of the Parc Monceau. This most exclusive corner of Paris emerged as the most egalitarian.

Certainly, I felt more at home there than at my school. I attended the École Chabrol, around the corner from the hotel. Though it was a working-class public school, I felt wildly inferior to my classmates, who arrived each morning in stylish pink or blue nylon smocks over their street clothes. My parents couldn't afford to buy me either. In my unfashionable dark woolen slacks and sweater from Cairo—and no colorful smock—I felt awkward and out of place. I suffered in silence, not daring to complain.

I thought yearningly back to my gray-and-white jumper with the embroidered crest, the uniform of the Lycée Français de Bab-el-Louk. My first day of school, I'd walked round and round the courtyard with my friends, feeling terribly stylish and grown-up in my elegant cotton dress. My books were in a brown leather satchel my father purchased for me and which I carried in my arms, *comme les grandes filles*—like the older girls.

On our last day on Malaka Nazli, as we pondered what remaining items to stuff in our suitcases, I kept staring at my leather bag, lying there on the dining room table. I started to reach for it a dozen times, but then Pouspous wandered in and commanded all my attention, so that the bag stayed on the table as we walked out the door.

I found the French girls of the École Chabrol alien so that even French, my native language, began sounding oddly foreign to my ears. Come lunchtime, I'd join my family at the Richer, even as the other little girls bonded in the cafeteria. During afternoon recess, I stood by myself in a corner of the schoolyard, praying for the games to end.

I am not sure whether I was pleased or mortified when my teacher surprised me with a gift.

One morning, I came to class and found a box on my chair. Inside was a small beige canvas book bag with leather trim inscribed "Pour Loulou." Had my kindly homeroom teacher noticed my discomfiture and sense of isolation? After months at the school, I hadn't made a single friend and barely spoke to anyone.

Somehow, I still managed to finish the year on a high note, collecting first prize at *la distribution des prix* held in the auditorium, where the mayor of Paris came to hand out the awards. I heard my name and stepped up to the stage, where stacks of books were lined up on a long table. Monsieur le Maire shook my hand as he gave me a set of books, tied together by a silk ribbon. Later that day, Mom and I trooped to the Cojasor to show off my bounty—the collected works of Hans Christian Andersen, a history book, and a fat Larousse dictionary. My mother was in a surprisingly jaunty mood, and Madame Dana beamed, thrilled either with my performance in school or, more likely, with the fact that Edith at last was smiling.

My birthday on September 19 was fast approaching. It would all be different then, I was sure of that. It was, in part, a child's mystical belief in the power and transcendence of birthdays. But it was also Cairo, where my birthday was a day when the world seemed to come to a full stop and all efforts were expended on making it as pleasurable as possible. My father saw to that.

It always began with the doorbell ringing early in the morning.

"Loulou, c'est pour toi," my father called out from his seat by the window facing Malaka Nazli. I'd run to the door to find Abdo, our porter, carrying an enormous white cardboard box bearing the distinctive name in blue lettering—Maison Groppi. Abdo would hand me the box, which was almost too heavy for me to carry. I didn't need to open it to know what was inside. The white box from Maison Groppi had been arriving on my birthday ever since I could remember.

To the end, my father had indulged my passion for Groppi's. Going there every afternoon was a ritual for the two of us, as it had been in former years for my siblings, and even my rebellious, wayward sister had happily accompanied Dad there and allowed him to treat her to one dessert after another.

Because we could never bear to leave, Dad and I would take a piece of Groppi's home with us—some of its delectable *crème Chantilly*. The cream had such a rich consistency, the staff simply placed it in a compact white cardboard box and tied a string around it, and voilà: it would withstand the taxi ride back to Malaka Nazli.

The box that I carried from the door to the living room was bigger. I tore it open and found what looked like a wedding cake, large, white, with pink sugar lettering and frosted flowers and curlicues and other elaborate decorations made of cream. On it were the words "Bon Anniversaire," in butter frosting. It was too early in the day to eat cake but no matter—it was also impossible to resist.

My father helped me cut large slices for members of my family, who gathered by the living room sofa. I was of course entitled to as much as I could possibly handle. I tried in vain to tempt Pouspous with some of my cake, holding it to her nose, but she seemed singularly unimpressed, the only resident of Malaka Nazli—possibly in all of Cairo—who had no use for Groppi's delicacies. My last birthday in Egypt was spent laughing, opening gifts, and hearing my family marvel at how Groppi's had outdone itself this year.

Paris seemed on the verge of obliterating all of that.

On September 19, I left school as usual to meet my family at the Richer for lunch. They were seated at their familiar table in the corner. The cafeteria was hot and more crowded on this autumn day, perhaps because of a new wave of refugees. The old countess came by and tried to pile extra portions on my plate. My family began eating as if it were any other day. I picked anxiously at my food.

It was when our aristocratic waitress returned to ask if I wanted more that I snapped. I took the plate heaped with hot food and hurled it across the floor, smashing it into dozens of small pieces that flew across the room, along with clumps of rice and meat and vegetables. My family sat, horrified at my uncharacteristic display of rage, aware that the entire dining room had turned to stare at us. Even the well-meaning countess was speechless, her gracious smile faded, and she looked as if she were about to cry.

I demanded to know where my cake was—my birthday cake from Groppi's:

"Mais où est-il?" I asked my bewildered parents and siblings. "Où est le gâteau d'anniversaire?"

My mother tried to level with me. It wasn't possible to get me a birthday cake this year, she said. It simply wasn't possible. Besides, I wasn't a child anymore. At seven, she said, I was old enough to understand how much our circumstances had changed: "Loulou, nous sommes à des milliers de kilomètres de chez Groppi"—Loulou, we are thousands of miles away from Groppi's.

My father rose slowly—it was difficult for him to get up from a chair these days. He motioned to me to follow him.

Together, we walked out of the Richer. I could feel all eyes on us. We strolled silently up the streets near Poissonière, passing countless fur wholesalers featuring luxuriant mink and sable coats in their windows—items that no one in the neighborhood could afford, produced for an outside clientele we never saw, who didn't deign to venture on to our street.

We continued walking until we came to a small bakery, the size of a broom closet, with a simple display of baguettes in the window.

Inside, a glass case featured a modest selection of cakes. They couldn't have been more than a couple of inches wide, but they were delicate and elegant in the manner of all French pastries, decorated with waves of frosting, cherries, strawberries, and puffs of Chantilly cream.

"Combien, madame?" my father politely asked the woman behind the counter.

She went through the price list with us. He finally pointed to a small cake with cream frosting and nodded. She placed it in a square white box, tied it with string, and handed it to my father, who in turn gave it to me to carry. I took the box, but I didn't say a word.

I had begun to feel horribly guilty and ashamed, as it finally dawned on me that my father had spent much of the family's allowance simply to get me to stop crying. It was then that I understood that life as I had known it was never going to be the same. While most children mature many years later, my coming-of-age took place on the afternoon of my seventh birthday, outside a small Parisian bakery.

CHAPTER 15

The English Lesson

One morning, we had an important appointment with HIAS. Before we left the hotel, my father insisted my two brothers join him in reciting the morning prayers. Both refused. "Sali, sali" (Pray, pray), he kept screaming at them, as Mom, Suzette, and I watched, paralyzed. By the time my brothers joined him and read—sullenly and by rote—more than an hour had passed, and we missed the meeting. HIAS officials were furious. If it ever happened again, they warned, we would be immediately cut off from any assistance.

Immigrating to the United States had turned into an ordeal, a painful, drawn-out process filled with bureaucratic land mines. While Israel, which took in all Jewish families, would have welcomed us, we had to persuade HIAS officials that we, in effect, *deserved* to live in America.

From the start, it was my father's job to engineer our entry.

French society was intensely patriarchal and accorded husbands and fathers an inordinate say in a family's affairs. We trooped along with him when he was asked to report to HIAS offices on rue Lota in the tony sixteenth arrondissement. We found ourselves in a neighborhood of vast, expensive tree-lined streets with private mansions and residential buildings set back behind tall gates. None of us said a word as we walked. We felt small and lost and thousands of miles removed from the grittiness of Montmartre and the rue du Faubourg Poissonnière.

This was the Paris, elegant and graceful, that we had expected to find when we'd left Egypt, the city Mom and I glimpsed when we went to Parc Monceau. But as with our visits there, we were being granted only an evanescent glance before being forced to turn back to our dilapidated quarters.

At HIAS, we were met by social workers who overwhelmed us with forms to fill out and barraged us with questions. Why were we so intent on moving to New York? they'd ask us again and again. Why had we left Egypt?

As if they didn't know.

My father told of the anti-Jewish sentiments we had witnessed toward the end, the fact that one by one all our relatives and friends had left, and even he, who felt an abiding love for the city and country of his youth, had come to realize our lives were in danger.

"I could no longer provide for my family," he told Mademoiselle Cygler, our caseworker; "my children had no future in Egypt." Unlike Madame Dana, her counterpart at Cojasor, the HIAS social worker seemed to have taken an intense dislike to all of us; she clearly resented our inability to make up our minds, our chronic unhappiness.

By the late fall of 1963, my father's efforts to get us approved for America had turned into a nightmare. HIAS and Cojasor seemed delighted with my older brother and sister, and were eager to let them emigrate. They were seen as exceptionally promising candidates. The problem was Dad. They were prepared to deny us all entrance because of their misgivings about him.

My father, the Cary Grant look-alike who had made beauties swoon and forced business rivals to heel, was deemed undesirable for America.

Too old, HIAS said. Too sick. Too infirm. Too beaten down. No prospects. Leon looked considerably older than he was in his weather-beaten raincoat and with the wooden cane he now relied on to walk. In the eight months since we had left Egypt, he seemed to have aged by almost as many years.

HIAS pointed out his limitations, the fact that he was so frail. It would be difficult, if not impossible, for him to find a job, they said. He would be unable to support the family, and then what would become of us?

"I worked until the day we left Egypt," my father coolly reminded them, seemingly unperturbed by the harsh put-downs. "I will work again."

He added: "Le bon Dieu est grand."

He wouldn't back down, and on that afternoon, the bureaucrats and social workers saw flashes of the indomitable will that had guided my dad through six decades, they caught a glimpse of the man of iron who had been born in one country, settled in a second, found himself exiled to a third, and was now determined—or perhaps merely resigned—to start life in a fourth.

But no matter what he said, the social workers kept reminding him of his physical limitations, the fact that he had trouble walking.

"I don't only walk—I can run," he cried out. And he looked as if he were suddenly going to bolt, wooden cane and all, out of the town house on the rue Lota and down the avenue Foch to the Champs-Elysées, and all the way back to . . . to where?

Anywhere but here, anywhere but this city that didn't want him to stay, yet had nowhere for him to go.

His most passionate arguments fell on deaf ears. In their internal deliberations, the resettlement czars were even more blunt in expressing their misgivings. Leon would never be a productive member of American society, they told each other in aerograms and telegrams that flew back and forth across the Atlantic. We were destined to be wards of the state because Dad wouldn't find work.

Suzette and César posed no such conundrum. They were energetic, well-spoken, healthy, and, above all, young—precisely what America wanted. All they lacked was the ability to speak the language. They had to start studying English immediately. We all had to—with the exception of my father.

Off we went to English class, with orders to learn to speak and read as rapidly as possible.

I was permitted to tag along. I was delighted: after months of feeling left out of adult decisions, I eagerly proclaimed my desire to take English lessons.

Classes, which were held close to the avenue Foch, were taught by a pretty American expatriate named Nancy Hakimian. Miss Hakimian was so perky and charming, César spent most of his time looking at her instead of paying attention to what she was teaching. My sister, always a diligent student, took careful notes. If Isaac came, I didn't see him. Mom, who went occasionally, stared forlornly out the window.

Miss Hakimian would begin each class with a dramatic stunt designed to get our attention: holding a large white porcelain mug high up in the air.

"Cup," she'd say.

We had to repeat after her: "Cup."

Then, in a feat that never ceased to amaze me, no matter how often I witnessed it, the teacher would take the cup and smash it in two.

"Broken cup," Miss Hakimian cried.

We were supposed to chant, "Broken cup," with the emphasis on "broken."

Those were my first English words—not "hello" or "my name is Loulou" or "good morning," but "cup" and "broken cup." The following session, she would begin the lesson by holding up the same cup and take us through the drill all over again.

The cup seemed to have magical regenerative powers—no matter how often it broke, it reappeared in one piece the next time.

As we advanced, Miss Hakimian expanded her repertoire of breakable objects to include saucers, plates, and glasses. She'd smash the saucer against the blackboard, fling a dish to the ground. They all made a comeback at the lesson that followed, and Miss Hakimian would smile mysteriously as if she alone knew the secret of their healing power. My English vocabulary grew to include "saucer" and "broken saucer," "plate" and "broken plate."

César, though beguiled by the lovely Miss Hakimian, seemed unable to grasp the most basic words or phrases. He was asked to take an aptitude test to determine what he should do when we reached New York. He was sixteen years old, and back in Cairo, he had studied to become an accountant, and hoped to go into business like Dad. The test was rigorous and took several hours to complete. In addition, he met with vocational counselors.

The verdict came in at last—César had technical skills, he was told, and was urged to pursue a career as an auto mechanic. My brother, who didn't have the least interest in cars, who couldn't even drive, didn't know whether to laugh or cry.

Unlike my employable siblings, Leon had no need for any English lessons. Even the bureaucrats marveled at his raffish British accent. They couldn't help wondering how he could converse in the King's English, while his children seemed ignorant of the most basic elements of the language. But they weren't prepared to cut him any slack. Convinced he would take from the system rather than contribute to it, they balked at letting him come to America.

At last, an obliging French doctor helped my father overcome a crucial hurdle. He certified that in his professional opinion, Leon was both healthy and fit to hold down a job. In form after form, the amiable young physician, Docteur Sananes, testified that Leon would indeed be able to work full-time, provided the employment was "sedentary." He downplayed the effects of Dad's broken leg and gave an upbeat, if not exactly rousing, endorsement.

Still, any celebration would have been premature; we had no idea when we would be checking out of the Violet Hotel. The quest to find us a home in America dragged on. Even after we heard we were approved in principle to settle there, the process seemed fraught with complications. HIAS launched a desperate search for relatives in the United States who would sponsor us, employ us, support us, welcome us into their homes, help us in any way. The cables crackled as officials contacted long-lost cousins from Brooklyn all the way to San Diego.

HIAS was encouraged that one relative had already stepped up to help—my Milanese cousin Salomone. Dad's nephew sent word he was prepared to give several hundred dollars to defray the costs of our move and had begun sending checks over to HIAS.

Our American cousins seemed to react altogether differently; admittedly, the initial requests from HIAS were ambitious. Could anyone subsidize our initial stay until we found our footing? The answer came back, a clear-cut no. Even a close family member like my mother's half sister Rosée complained that she and her children in Brooklyn were barely making ends meet.

HIAS proceeded to ask whether these relatives would take us in, allow us to live under the same roof for some weeks or months. Again, the answer was a resounding no. There was no room in anyone's cramped American dwelling for my family.

The agency drastically scaled back its demands. Would our relatives in New York at least help look for a suitable apartment for us? Once more, the answer came back no. My mother's relatives said they would try, but were too busy to find us a place to live.

HIAS made one other intensely modest request. Would someone, anyone, come greet my family at the pier?

There was no reply.

News of their balkiness reached my mom, who was thoroughly wounded. She had always adored her half sister Rosée, and spoke fondly of how warm and welcoming she and her children had been back in Cairo. What had happened to make them so distant and self-absorbed—so emotionally stingy? Was that one of the dangers of becoming an American?

Perhaps that was another English lesson, one that involved not shattered dishes but familial bonds that were irrevocably broken.

WE STILL HADN'T UNPACKED by the early winter of 1963. Whatever progress the resettlement agencies were making on our behalf seemed painfully slow. My sister, in particular, despised the lowly secretarial job she had finally landed at a nearby textile shop, which was off the books. One day, she discovered she had misplaced an entire month's salary: the money had been either lost or stolen. The episode only underscored her sense of futility.

A local furrier hired my younger brother, Isaac, now thirteen, to work with him in his small factory near the Violet Hotel. César, meanwhile, was racking up tips doing odd jobs, running errands and delivering packages for a fabrics store situated on the passage Violet. Its elderly Romanian owner developed a fondness for my oldest brother and dispatched him all over Paris to deliver bolts of the fine wool he imported from England. Generous tips made it easier for César to enjoy his nocturnal escapades with his new friends, other teenage refugees staying in nearby hotels. His greatest joy was when he received a shiny five-franc coin from wealthy clients—two francs more than the daily allowance allotted for each of us by the Cojasor. The large tips only intensified his sense of having arrived at a city of endless possibilities.

Late at night, César and his friends would amble over to lively Montmartre, where the cafés were always open and welcoming. At a bar near the lobby of a local hotel, he noticed two elegant, intriguing-looking women in heavy makeup and expensive clothes, night after night having drinks together. It took a while before my brother and his friends realized that the chic women were men, and they understood they were a million miles from Cairo.

Near the Violet Hotel was another landmark, the Folies Bergère, the music hall that was as iconic a symbol of Paris as the Eiffel Tower or the Louvre, and whose gorgeous showgirls were renowned the world over for their sex appeal and glamour. César took to meeting his buddies by the Folies Bergère in the evenings. They'd stand on the corner in their *blousons noirs*, trying to look suave as they smoked Gauloises and eyed the dancers sauntering in and out. Occasionally they had a gig inside. The manager had spotted the youths loitering by the theater, and he hired them on some nights to prime the audience. While enjoying front-row seats, my brother and his friends were paid to clap loudly and cheer as the showgirls performed their numbers. It was delightful work, a far cry from the dismal routine of a refugee, and it also gave César a window into the mysteries of the legendary showgirls. Onstage, they sparkled and smiled and performed with exceptional grace and agility. But up close, my brother noticed with dismay, they were a lot older, more jaded—more ordinary—than he could ever have imagined. Stripped of their glittering costumes, the beauties of the Folies Bergère weren't even that beautiful.

If he were more introspective, César could have viewed the experience as a metaphor—for Paris and women and life beyond Cairo. He could have pondered how beauty invariably disappoints, and how nothing in this world, not even the most dazzling city and its most delectable women, ever lives up to expectations.

César in his blouson noir *(black leather jacket), Paris, 1963.*

But with his extreme literal-mindedness, my brother walked away from his work as a shill for the Folies Bergère only with a vague distaste for showgirls.

One night, as he stood with his friends eyeing the crowds gathering in front of the dance hall, he spotted a tall man in a dark luxuriant wool coat and top hat, walking out the door with a beautiful woman clinging to his arm. He recognized him at once. It was Maurice Chevalier, the movie star who was as much a symbol of Paris as the Folies Bergère. But unlike the showgirls, Maurice Chevalier didn't disappoint a bit. He cut a striking figure, as dapper and distinguished as in his movies. My brother and his friends could only stare as he flashed his famous smile their way, doffed his hat, and kept on walking. César had again the sensation of living in a dream city, and whenever he'd think about our time as refugees and how we had survived and what he had loved the most about Paris, he would conjure up the night he saw Maurice Chevalier.

My mother didn't care for any of this.

She was deeply disturbed that her eldest son was staying out all hours of the night and languishing on street corners like a hoodlum. She was the first to realize, even before my father, that she and Leon no longer exerted the same power over their children as before. In the world outside of Egypt, my siblings were either too alienated or too rebellious to heed what my parents had to say.

She turned to the social workers at the Cojasor for help, and they duly recorded her sense of desperation in Dossier #45,135. But Madame Dana and her colleagues confessed there was nothing much they could do. We were in a state of limbo. There were restrictions about working full-time, and it was impossible for César or Suzette to go to university because we could leave France any day. The social worker could only counsel patience.

Meanwhile, I discovered my favorite spot in all of Paris. It was a small doll factory a few steps up the passage, whose door was usually left ajar, so I could peek in every morning on my way to school and on my way back. The factory was heaven as defined by a seven-year-old girl—hundreds and hundreds of dolls, in various states of completion and undress, all lined up on shelves.

There were dolls without heads, and heads without dolls. There were dolls on racks with no clothes on, and dolls decked out in their full finery. There were dolls with long cascading hair and dolls that were bald. On a stand, miniature wigs were piled one on top of the other— blond curls, red tresses, dark sweeping chignons, sultry pageboys, all waiting for their turn to be placed on the head of some lucky doll and make her beautiful. In those restless final weeks in Paris, that became my favorite activity: marching obsessively past the factory's open door and staring at the dolls.

A breakthrough came in late October: my father was summoned to the rue Lota and asked by HIAS to sign a promissory note.

We were going to America. The document stipulated that we would have to repay any money advanced to us to cover the expense of traveling by ship to New York, including taxes and inland transportation. There was also the freight cost of moving our 1,510 pounds of luggage— a staggering amount of pajamas, lingerie, bedding, pots and pans, sardine cans, and one twenty-year-old wedding gown.

The HIAS loan was routine for the agency, which after all was in the business of sending destitute refugees to America and other countries. The agency would typically purchase tickets and advance families the money to cover travel expenses, expecting them to repay the agency years down the road, when they were back on their feet again.

No amount was specified on the promissory note Dad was asked to sign. We were agreeing to a loan for an unknown sum. My father signed it anyway, of course, though it was only months later, in New York, that we learned the extent of our indebtedness: $1,199.94.

We were finally on our way when I suddenly developed another of my mysterious maladies. I became violently ill with a fever, a rash, a stomachache, and an excruciatingly sore throat. My mother fretted as she bundled me up in bed under every blanket she could cull from the Violet Hotel's poorly stocked linen closet.

My father decided to summon the kindly Dr. Sananes, who had given him a glowing bill of health: he would pay for the house call using a week's allowance, if need be, but at least I would be seen by a proper physician, without having to trek to one of the public clinics, where we had found the care to be singularly unimpressive. In my case, all that our local *dispensaire* did was order more and more blood tests. Typically, I wasn't even seen by a doctor but by nurses. I longed for the distinguished men in white coats bending down to examine me in their private studies in Cairo. I even missed the Professor, of the white gloves and the cold and formidable manner.

Did the West really have a superior medical system to ours? Not in the Paris we had come to know.

When Dr. Sananes arrived, he glanced at our shambles of a hotel room, with suitcases piled one on top of the other, and shook his head. After examining me, looking at my throat, taking my temperature, and peering at the rash I had developed, he rendered his diagnosis.

I had *la scarlatine*—scarlet fever—a dangerous infection whose key symptom was the pink rash that seemed to be spreading. The disease often began as a simple sore throat, he explained, but could turn fatal. We shouldn't even think of traveling to America now—not for months. He offered to sign a note asserting that under no circumstances could we leave France.

My parents were stunned. How on earth had Loulou contracted scarlet fever? they asked each other.

News of my illness reached HIAS, the Cojasor, and every other agency handling our case. The officials reacted with alarm. The diagnosis was so dire, it threatened to unhinge their meticulous travel plans. HIAS had finally managed to book the family tickets aboard the *Queen Mary*, which was to sail from Cherbourg to New York in early November. Clearly, because of my scarlet fever, we weren't going to make the voyage.

Telegrams about my plight were dispatched to any number of overseas offices. "The youngest, a little auburn-haired girl, is sick," read one cable. There was now an even bigger question mark. When would my family, which had already consumed an inordinate amount of resources, attention, and psychic energy from the local and international relief agencies, finally be able to leave France?

Between our inability to make decisions, our endless need for services, my siblings' unhappiness, and the clear signs of discord in our family, we had taxed these relief and support agencies to their limit, and they were anxious for us to be on our way.

Agency officials debated how it was possible for a seven-year-old girl in a Paris hotel to contract *la scarlatine*. Was this an elaborate plot to remain in France? They had certainly heard of refugees delaying their departure to linger in Paris, but inventing a deathly illness for a young child would be a first.

At last, agency officials thought of sending over a seasoned doctor to examine me and confirm the diagnosis. My mom was overjoyed. "Le bon docteur arrive," she cried, once again praying for the mythical, all-knowing physician of her imagination.

For once, she wasn't disappointed. A distinguished French doctor holding a small black bag knocked on our door and made his way immediately to my bed. He seemed oblivious to the chaos in the room. After examining me thoroughly, he rendered his diagnosis: I only had a sore throat, maybe a severe sore throat, at most a strep.

What of the scarlet fever? my father asked.

"Absolument pas," he declared. With his calm, commanding manner, he seemed to have stepped out of my mother's dream of *le bon docteur.* He scribbled a prescription for an antibiotic on a pad and ordered me to stay in bed for another week. After that, he said, I was free to go to school—or even to America.

When my dad offered to pay him, the doctor said, "Non, merci," shook hands, bowed, and left.

My father was so distraught at Dr. Sananes and his dire pronouncement that he called him. Why had he frightened us with his diagnosis of scarlet fever, he asked the young doctor, when all I had was a sore throat? The physician seemed puzzled. He had assumed we *wanted* to stay in Paris a while longer, that we needed more time to prepare for our journey. He had tried to do us a favor now by stressing the grave nature of my illness, knowing that a family with a sick child would be able to stretch out their stay in France.

Within a matter of days, HIAS announced they'd bought new tickets for us on the ship's next crossing, in early December. The ocean liner was to sail from Cherbourg, with a weeklong voyage that would get us into America shortly before Christmas. They were purchasing five and a half tickets for the family.

I was the half.

There were no elaborate preparations for our voyage this time, no major shopping expeditions. In one brief burst of anxiety, my mother insisted that my father take me at once to a shoe store. I had to have a pair of boots, she said: it was the only way to shield "pauvre Loulou" from a country likely to be even colder than France.

Hand in hand, Leon and I walked to Montmartre and its bargain-basement stores. There, in the window of a popular discount chain, were dozens of children's boots—boots with thick fur lining, suede boots, rubber boots, and boots entirely of leather. I had never owned a pair of boots, so the mere act of trying them on felt like an adventure.

I pounced on a pair of jaunty galoshes. My father agreed to buy them, though they were plastic and flimsy. But he also nudged me gently toward a practical pair of blue suede boots that reached past my ankles, with furry pile lining and thick yellow soles of caoutchouc—rubber. They looked like an Eskimo could happily have worn them. My father examined the lining with an expert eye and nodded his approval.

I felt tough and invincible as I stomped around the store in my arctic boots. I felt ready for America.

A couple of weeks before we were scheduled to leave, we heard a scream coming from the passage Violet.

"Ils ont assassiné votre président!" the porter was shouting toward our window—They have killed your president! My family looked at one another, thoroughly befuddled. Had Nasser been murdered in Egypt? Had they assassinated King Farouk in his Italian exile? Or was it General de Gaulle who had been killed here in Paris?

It took a few minutes before we realized that "our president" was the president of the United States. John F. Kennedy was dead. All day, my family huddled around the small table. The small leather-cased transistor radio we had purchased in Alexandria, and which had been our lifeline to the outside world since leaving Egypt, was blasting.

I caught only the same six words, "Le Président Kennedy a été assassiné," repeated over and over again. My parents and siblings seemed shaken. They spoke in such low voices, I couldn't make out what they were saying.

Discussions about where we should go resumed with intensity, and were more agitated than ever. Should we really move to New York? What kind of country were we going to that murdered its own leader? Even Farouk, the victim of a military coup, had been permitted to leave safely and sail out of Egypt aboard the *Mahrousa*, the royal yacht.

Our decision appeared terribly flawed, though we felt helpless to change course. Of all of us, my sister reacted the most emotionally. Only when Alexandra died had I seen her cry so copiously. To Suzette, the murder of JFK underscored the inchoate fears and misgivings she'd felt all along about the family moving to America, the sense that it was fundamentally the wrong place for us.

A couple of weeks later, we boarded the train to Cherbourg, where the *Queen Mary* was waiting to take us to America. We were all glum. César, who had liked Paris more than any of us, felt as if he were waking up from a dream.

When we arrived in Cherbourg, my father and I broke off from the rest of the family and took a long walk. It was after sunset, and we always loved to walk together at night, though I noticed that his gait was more tentative in the dark, and I wondered if he was in pain.

Dad and I found ourselves standing in front of a massive ocean liner shimmering in the still dark waters. It was the *Queen Mary,* so close we could almost touch it. Vast and imposing and stretching out for what seemed like a mile, it was like no ship we had ever seen before. In comparison, the *Massaglia* was a shabby little rowboat. Yet despite its majesty and heft, the *Queen Mary* offered us little comfort, and indeed, even heightened our sense of terror.

Or perhaps it was simply despair we felt at finding ourselves staring at another ship we would be boarding for yet another voyage into the unknown. I held my father's hand a little more tightly as we lingered, dazzled and scared at the same time.

Dad was still in his faded raincoat, which had become his armor during our months in Paris. He was trying to come to terms with the fateful decision he had made. He realized, of course, that choosing America over Israel dashed any hope he had of rebuilding what he had lost. Never again would he live within walking distance of his brothers and sisters, never again would the family come together as in the dining room on Malaka Nazli, the men in their crisp new cotton pajamas, the women in their elegant robes, looking to him—the Captain, the patriarch—for guidance.

My father's entire life had been guided by the primacy of family that was the Aleppo way. Family above work, family above money, family above ambition, family—though his wife would deny it—above personal pleasure. Yet there we were, on our way to a city where we had no one except a handful of relatives who didn't care enough about us to meet us at the dock.

CHAPTER 16

The Wrath
of Sylvia Kirschner

My jaunty gray Cicurel coat did little to shield me from the arctic chill of Pier 90, where the *Queen Mary* berthed after arriving in New York. We stood on the ship's bow, looking down toward the dock. The entire landscape was white, and we were mystified. César decided to investigate.

"What is that on the ground?" my brother asked the passenger next to him.

The man's eyes widened, as if he had encountered a Martian.

"Snow," he replied, and then edged away.

After going through customs, we stood in front of the pier, our mountain of luggage piled up around us, waiting I wasn't sure for what. Our fellow travelers were leaving us behind one by one. They disappeared into waiting arms or waiting cars or waiting taxicabs even as we continued to stand out in the cold. We had come so far, yet we didn't know where we were going, and we had no one to take us there.

We were somewhat in shock, staring at the cars bobbing up and down the West Side Highway; they were all so enormous—so outsize:

nothing like the endearing little Citroëns and Renaults we were so used to seeing on the streets of Paris. Those were my first impressions of America: the bitter cold and the large imperious automobiles that occasionally came to a full stop and picked someone up, but never me or my family.

I went over to my father and reached for his hand. He had on his thin old raincoat, which was, if possible, even flimsier than my woolen coat with its matching scarf, but he didn't complain, though I noticed that he wasn't wearing gloves, and his hand felt like ice. He was strangely silent: no cries of "Ragaouna Masr." He simply stared, as we all did, at the grayness of the sky, the whiteness of the ground, the bleak horizon of low-lying buildings and the cars moving, moving along the highway.

I tugged at his sleeve, which I did whenever I wanted his attention, and he managed to dig into his pocket and remove a piece of candy. He had bonbons left over from the stash he'd collected on the *Queen Mary*, where every night was a feast, an occasion to shower passengers with desserts and favors and music and treats.

Our passage across the Atlantic had felt like a holiday cruise, one long luxuriant party. The sheer opulence of it all left us almost in a daze after the miseries and privations of the prior year. It was sheer luck that we had maneuvered a trip on a grand ocean liner, instead of flying coach for ten hours or more on Pan Am, the usual mode of travel for refugees. My father told HIAS that he couldn't tolerate an extended plane ride because of his leg. We were steered toward the *Queen Mary*, since its departure for America coincided with the date HIAS had decreed we should leave. Before we knew it, we had tickets on the grandest ship afloat, fit for dukes and duchesses and debonair film stars.

Admittedly, we had the least expensive accommodations available—third class, modest quarters for the most budget-conscious travelers. They didn't strike me as particularly modest, though, not compared to our recent digs—the cabin by the engine on the *Massaglia,* or the Violet Hotel.

The gracious, exquisitely polite culture of the *Queen Mary* suited us to perfection: at last, a world outside of Cairo where people weren't rude or impatient, where they were actually solicitous, and deigned to show us some kindness and concern.

My father felt at home with the British crew. He bantered amiably with everyone from the captain to the purser, showing off his command of the language and his exquisite accent. None of us could compete with him when it came to speaking English.

We sailed a couple of weeks before Christmas, and a holiday mood prevailed. There were nonstop diversions—concerts and dances, movies, plays, games, and soirees, organized by an energetic crew that seemed interested in our well-being—making certain we were happy and enjoying ourselves.

I couldn't remember the last time anyone had cared whether we were happy.

While I stuck close to my dad, my siblings roamed a ship that felt as vast as a city. Though first class was technically off-limits, César made friends who let him peek at its dazzling ballrooms and lounges, tall staircases and elegant carpeted suites. At night, he went dancing at the clubs that catered to teenagers and featured the latest American hits, including a jazzed-up, souped-up Latin version of "If I Had a Hammer," sung by Trini Lopez.

After months of the greasy dishes of Le Richer, dining on the *Queen Mary* was the greatest extravagance of all. We enjoyed gourmet meals served by our personal waiter who boasted a command of some twenty-five languages, and had each language that he spoke stitched into the fabric of his sleeve. Every lunch and dinner, he appeared magically at our side, offering to translate the menu into the language of our choice. Unlike the lone other family on our side of the ship who kept kosher, and insisted on Yiddish translations of every entrée and appetizer, we remained mostly silent and tried to follow along in English. We ate on elegant porcelain china, using fine silverware engraved with the word *Kasher.*

I sat, as always, at my father's side; he was more cheerful than I'd seen him in months. It was as if the magical powers of the *Queen Mary,* its British culture, its deferential staff, its soothing vegetable broth prepared in the ship's kosher kitchen, made him feel hopeful for the first time about our lives outside of Egypt.

We had felt protected the entire time we were aboard the ship, but now, on the pier, the old feeling of being lost and at the mercy of an uncertain fate returned. I noticed that my parents and siblings kept looking anxiously out toward the highway, as if some familiar face would materialize from the icy gray blur. No one could explain to me what we were doing, why we had come all this way only to be left out in the cold.

We were officially welcomed to America by an HIAS bureaucrat, who apologized profusely for being late. She handed my father $50 to help tide us over those first few days, and arranged for a taxi to transport us and our suitcases to our hotel.

The Broadway Central was a lumbering old hotel, long past its prime, perched between Greenwich Village and the Bowery. It had once housed any number of illustrious guests and visitors, from Diamond Jim Brady to James Fisk, the railroad tycoon who was shot there, to Leon Trotsky, who waited tables before hastening back to Russia to lead the Red Army.

But now, in the early 1960s, it was so down-at-the-heels it catered mostly to needy low-income families and stray out-of-towners and refugees like us who couldn't afford any better—a forerunner of the welfare hotels that would become commonplace.

Though we were given a suite, our accommodations were, if possible, even more squalid than at the Violet Hotel. We had a small kitchenette and two large drafty rooms, where beds were lined up one next to the other as in a hospital ward. There were only five beds for the six of us, so I doubled up with my mom in a small bed, close to a wall with a large gaping hole.

We were used to balmy winters, and even Paris had been mild the year we were there, but here it was freezing cold. I went to bed every night in my street clothes—a pair of gray wool slacks from Cairo and a turtleneck sweater.

My mother thought I was being silly, as did the rest of the family. No one could understand why I insisted on sleeping in scratchy woolen street clothes instead of the soft and toasty flannel pajamas they had managed to retrieve from one of the suitcases, and I am not sure I understood myself.

We fell back into the nerve-racking rituals of people with nothing to do. I took walks with various members of my family—slow walks with my father, who was in constant pain, aggravated by the frigid temperatures; brisk walks with César, who was curious about America but not in love with it the way he had been in love with Paris; anxious walks with my mom, who seemed bewildered by the Village and New York in general; quiet walks with my sister, who took me again and again to Washington Square Park.

On the benches were people clad entirely in black, who looked like no one I had ever seen before. I couldn't help staring at these strange creatures seated amid the snowy white splendor of Washington Square Park. "Ce sont des bohémiens, des 'beatniks,'" my sister explained. We'd sit in one of the benches and stare at them, hoping they would approach us. They had eyes only for each other, and neither I nor my sister, all bundled up in our layers of Mediterranean garb, could possibly be part of any group. We were still outsiders, even to the beatniks, the quintessential outsiders.

When she walked alone, Suzette would occasionally find herself accosted by a beatnik, asking her for a handout. She'd shake her head no and continue walking. But she felt strangely guilty about turning them down, though she had even less than they did.

More inviting even than Washington Square Park was our local supermarket. I'd never been inside a supermarket before, and I found it dazzling, especially the fruits and vegetables, which I was used to buying loose by the pound, in outdoor stalls or from the vendors who roamed around Malaka Nazli. Here, they came packaged in green paper cartons, tightly wrapped in a layer of cellophane, so that even ordinary grapes or pears seemed remote and shiny and untouchable. I wondered why anyone would take the trouble to cover bananas or green beans in plastic, when anywhere else in the world, it was possible to simply reach for some. That must be America, I decided: a country where even commonplace items like apples sparkled and looked expensive and desirable beneath their plastic sheathing.

Bread was another mystery. I was used to tall thin golden baguettes purchased fresh from bakeries all over Paris, and in Cairo, we enjoyed hot round pita bread that came from the oven. But here, the package of white bread looked nothing like the bread I knew. It was all dough with practically no crust, while I was used to crust and very little dough.

We eyed the packages of Wonder suspiciously, inspecting them closely.

I was anxious to sample some, but my father seemed horrified: "Loulou, ce n'est pas du pain, ça," he said; This is not bread. We never bought white bread from the supermarket near the Broadway Central and rarely, if ever, later on.

A few days after we'd arrived, the resettlement agency called, asking to see us. HIAS had discharged us from its files; our only remaining contact involved the debt we had incurred for the tickets to sail aboard the *Queen Mary,* and which my father had agreed to repay over time. Now we were in the care of NYANA, the New York Association for New Americans. Mom, who loved to Frenchify every English name, promptly dubbed it *la Nyana.*

My father and César made their way to the agency's lower Manhattan office to meet with the social worker in charge of our Americanization. Sylvia Kirschner, a tough-talking veteran, seemed from the start to take an active, almost visceral dislike to my father. She offered so much advice it was dizzying. Our stay at the Broadway Central had to be as brief as possible. The family needed to find a place to live. My dad, my older siblings, and even my mom all had to go out and find work. We had to master English and meet people and make friends and lead normal lives again.

The initial meeting had the feel of a police interrogation. Why hadn't we begun to look for an apartment? Where were my mother and the other children? Why hadn't they come, too? Had we made any contact with relatives who could help us find work or a place to live? We had been in America all of five days; Mrs. Kirschner seemed in an awful hurry.

My father sat there, listening politely, talking only when she lobbed questions his way. He was so quiet and deferential that the social worker misunderstood—the way that she would consistently misunderstand him. She mistook his silence for contempt, and decided he was being obsequious when he was simply trying to be gentlemanly, more so than usual because he knew that this woman held our fate in her hands.

Unwittingly, Dad had incurred the wrath of Sylvia Kirschner.

It wasn't that Mrs. Kirschner was blind to my father's frailties—his advancing age, his deepening infirmities, his growing dependency. On the contrary, in page after page of notes that read almost like a diary, she chronicled my dad's failing strength, observing that he "looks considerably older than his age, walks with a pronounced limp and also very slowly due to his leg fracture," and "was obviously in pain." Even in the relative comfort of her office, she noticed that he could barely sit still without shifting his leg or grimacing, and he was so "very tired."

Yet, faced with a man clearly in decline, Mrs. Kirschner seemed unmoved. She found him troubling. Though skilled and vastly experienced, a professional who'd helped thousands of immigrants make the transition from the old world, making that transition had been based on the act of *letting go*—abandoning belief systems that were quaint and out of date in favor of the modern, the new, the progressive ideas that were so uniquely American.

That is what assimilation was all about, yet the overly polite gentleman with the vaguely British accent and the severe limp rejected the notion out of hand.

My father was by no means convinced the values of New York trumped those of Cairo. He couldn't see abandoning a culture he loved and trusted in favor of one he barely knew, and which he instinctively disliked. He preferred being an old Egyptian to a new American. He had, in short, no desire whatsoever to assimilate. "We are Arab, madame," he told Mrs. Kirschner.

It was a tragic clash of cultures and personalities. Both strong-willed people, my father and Sylvia Kirschner were set in their ways, and adhered to belief systems that were worlds apart and could never, ever be reconciled. Like boxers in a ring, they stood in their respective corners, determined to fight to the final bell for the principles they cherished.

And in a way, the test of wills between Sylvia Kirschner and Leon Lagnado in a small refugee agency in early 1964 presaged the conflicts my family would face for years to come in America, where our values and feelings about the importance of God and family and the role of women would constantly collide with those of our American friends. It also hinted at the larger, more terrifying and far deadlier conflict that would break out between the United States and the Muslim world decades later, when the United States would seek to spread its belief in freedom and equality only to find itself spurned at every turn by cultures that viewed America as a godless and profoundly immoral society.

Leon could have been a criminal, a jewel thief, a philanderer, a swindler: nothing could have offended our social worker more than his refusal to conform and change and cast aside those values she clearly viewed as virtually un-American and utterly repugnant.

In her eyes, my father was a patriarch in a land where there were no patriarchs. He wanted to rule over his wife and children—perhaps even his social worker—even though men weren't supposed to do that anymore. "He is an extremely rigid person, with limited horizons, has an Oriental psychology, covered up by a veneer of manners," she wrote. My dad and his views were hopelessly at odds with the enlightened society he had been fortunate to enter.

Or maybe not so fortunate. In one of her more insightful moments, Mrs. Kirschner remarked that my father "regards the immigration as a calamity rather than as an opportunity."

Barely a week later, the six of us trooped down to the tip of Lower Manhattan to meet with the redoubtable Mrs. Kirschner.

She looked us up and down, taking notes, then came to me, peered closely my way, and took more notes. The only one she approved of unreservedly was Suzette. From the start, the two laughed and chatted as if they were old friends. My sister turned on the charm. "A very attractive, articulate young lady," Mrs. Kirschner raved in her case files.

Not all of us fared as well.

César seemed to annoy her almost as much as my father. She didn't accuse my teenage brother of being old-world; she simply resented a sense of ambition she felt went beyond his natural abilities. Mrs. Kirschner seemed troubled by my older brother's outsize dreams, the fact he resisted taking an entry-level job as a messenger or a clerk. She stressed the need for him to be practical and start working.

Over the years, my brother would blame Sylvia Kirschner and *la Nyana* for the path he had taken, for the fact he had gone to work at eighteen, stuck in a series of menial and low-paying jobs, when he should have been attending school and building his career.

Instead, because of the fateful decree that he land a job, any job, the college degree that César could have earned in four years took him a decade to complete. He had no choice but to attend night school, where most students were immigrants like himself, which only underscored his feelings of apartness and alienation. His master's degree, which should have taken two years, took five instead. By the time he was done, César was thirty-five.

Mrs. Kirschner was deeply sympathetic to my mother and anxious to help her, to change her, to help her take advantage of the opportunities that had been denied to her as a woman in Egypt.

Mrs. Kirschner became obsessed with my mother's appearance, the fact that she was toothless and looked older than she was. The idea that a forty-two-year-old woman would walk around without any teeth struck her as almost barbaric. In the social worker's eyes, Edith was timid, quiet, anxious, and clearly under my father's spell. Mom "gave the impression of a frightened person," the social worker wrote, "emphasized by her enormous black eyes which stare almost childlike for protection." Leon was to blame. All the conflicts and problems and pathology she saw in my family were largely the result of his impossibly domineering personality.

What could America do for such a woman? It might not be able to give back her self-esteem, but it could at least provide her with a set of false teeth.

After grilling both my parents on why Mom hadn't seen a dentist in Egypt, she ordered her to go immediately to a dental clinic to be fitted for a set of dentures; the agency would foot the bill, Mrs. Kirschner grandly decreed, along with her edict that we had to leave the Broadway Central.

Most of the other refugees from the Levant had landed in one small corner of southern Brooklyn. My family was so unmoored, it made eminent sense for us to rejoin our lost community. César and my father journeyed daily to Bensonhurst, the ten-block area where refugees from Cairo and Alexandria had fetched up in an urban encampment of low-lying redbrick tenement buildings and simple two-family homes. They walked in the bitter cold, searching and searching for signs in the window proclaiming "Apartment to Let."

Early on, they stumbled on one promising prospect—a small

apartment on the second floor of a house owned by a dentist, Dr. Cohen. My father engaged the dentist in conversation, hoping to negotiate a more affordable rent by stressing his deep commitment to Judaism, his habit of attending services every morning. Only at the end did Dr. Cohen blurt out that he wasn't Jewish. Dad, stunned and confused, left, bewildered by this land where nothing was what it seemed, not even a doctor named Cohen.

My mother and I also tried our luck. We ventured out to Brooklyn and wandered around, on the lookout for To Let signs. Most were beyond our means. Exhausted, we decided to pay a call on my mom's relatives and the stepsister she hadn't seen in five years, not since Tante Rosée and her brood had left Egypt.

I had never met Rosée, but my mom had spoken worshipfully of this older woman who had been like a mother to her. She loved to recall her engagement, when Rosée took it upon herself to sew her bridal gown. At the end, Rosée had tucked strands of her own hair inside the hem for good luck. My mother could never part with it, and there it was, lying in one of the twenty-six suitcases that had followed us to America.

My aunt lived on a staid two-way street of shops with apartments upstairs. It was the Christmas season, and the area glistened with holiday decorations. What astonished me most wasn't the abundance of trees and lights and plastic reindeers but the lone Hanukkah menorahs with their soft orange glow on display in so many windows. I had never seen an electric menorah before; at home, we lit small wicks that floated in a pool of oil and water.

I was used to a culture where religion was practiced discreetly, behind the closed doors of one's house or synagogue, yet here were Jews observing their holiday as openly and assertively as their Catholic neighbors with their wreaths and garlands and Merry Christmas signs.

Tante Rosée and my mother embraced, and the obligatory *café Turque* was brought out on a tray, but later, Rosée began to lecture Mom on the etiquette of visiting in America.

New York wasn't like Cairo, she declared. The custom of dropping in on friends or even relatives without prior arrangements simply wouldn't do. "Here, in America, you have to call first," she told my mom. I noticed my mother freeze, then smile blankly.

We wandered back into the street, but suddenly, the glimmer of the holiday lights seemed a lot less hopeful. We felt far away from Malaka Nazli and the stream of relatives and friends who dropped in on us constantly, and on the long trip back to Manhattan, we were both silent. It had begun to dawn on us that this culture we were being asked to embrace, with its promise of riches and opportunity, could be as savage as the December night air that pierced our flimsy Cicurel garb.

The tough, forbidding woman we had met this evening bore little resemblance to the Tante Rosée my mom remembered from Egypt. She preferred to think of the last exchange as an aberration, a mistake, and remember her stepsister as the person who had taken it upon herself to sew her a magnificent wedding dress and had brought her back to life as she struggled with typhoid and the loss of her blue-eyed baby girl.

IT WAS CERTAINLY THE coldest winter we had ever known, but it was also one of the coldest winters New York had ever known.

Every day, César and my father trudged arm in arm through the streets of Brooklyn, humbled by the snowdrifts that were several feet high. My dad's limp grew worse, aggravated by the perilous walks on snow and ice. My brother wasn't faring well either. He was so thin—nearly six feet tall, he weighed only 140 pounds and suffered from a terrible cough, the result of the frigid weather and his habit of smoking several packs of cigarettes a day.

Mrs. Kirschner suggested he and my father go immediately to the Northern Dispensary, a clinic in the Village.

To help César recover, the agency also approved the purchase of a winter coat. Together with my dad—the two had become inseparable—my brother set out for S. Klein's, the discount department store in Union Square. There was a sale, and out of a combination of prudence and panic and confusion, my brother selected an overcoat that was nearly ten sizes too big.

The days of the sleek, fitted black leather blouson were over. The dark woolen coat on sale for $17.50 was size 46—suitable for a man several inches taller and many pounds heavier. It reached past his knees, the sleeves were way too long, and the shoulders drooped, so that its style was raglan.

When César modeled his new coat, my father nodded his approval and remarked that my brother would grow into it. It would surely help him survive his first American winter. Alas, the opposite proved to be true. The coat was so large it shielded him far less effectively than one his own size.

It was as if, marooned in America, we had lost our perspective, our sense of proportion. My brother, who had always liked well-tailored, fashionable clothing, ended up purchasing a coat that was neither. My father, who had paid such meticulous attention to cut and style, was now unable to look at a garment his eldest son was buying and point out its obvious flaws.

Worse still, Dad had become oblivious to his own appearance. For the first time in his life, he was dressing sloppily, and paying almost no attention to the way he looked. Mrs. Kirschner was struck by my father's battered, impoverished garb. Once the essence of style as he ambled through Cairo in his immaculate white suits, he was described in her notes as "shabbily dressed." What about all the clothes in the twenty-six suitcases? she wondered.

My father was silent, both when he sat in Sylvia Kirschner's office and back at the Broadway Central. He had always kept his own counsel, and he wasn't about to start confiding in her or us or anyone his despair over finding himself stranded in a hotel room in Greenwich Village in winter, with no means and no prospects and a little girl asking for white bread and fruit wrapped in cellophane.

And so the only sign of his inner struggle was in his clothes—frayed, careworn, slightly askew.

One morning, we woke up to the sound of clanging bells. We could hear people shuffling in the hallway, then more bells. It was barely five o'clock—we had no idea what was going on. We were only days before Christmas, yet my mother sat up and cried out delightedly, "Ce sont les cloches de Pâques"; They must be Easter bells. There was furious banging outside our room, and then the cry, "Fire!"

The Broadway Central was in flames. We had to evacuate immediately.

Since I had gone to sleep in my usual gear of wool slacks and a sweater, I sprung out of bed, fully dressed. But no one else was ready. Everyone seemed either frozen, unable to move, or scurrying around in a state of panic. My mother still couldn't believe what she'd heard was a fire alarm. My sister fretted about what to wear. My father moved more slowly than usual, unsure what to take with him: Precious papers? One key suitcase among the twenty-six? César recovered his composure and grabbed his wallet along with odds and ends—travel papers, photos of childhood friends, a few American dollars, and, for good measure, a couple of Egyptian pounds.

I kept yelling out to everyone, "Allons, allons," Let's go, let's go, my survival skills finely honed even at age seven. There was smoke and pandemonium in the hallways, and people in bathrobes and hair curlers were crying and hurrying toward the stairs and elevators.

Finally, after what seemed like ages, we were all ready, except my sister, who was still fussing by the closet. My father shuffled out, taking nothing except his wallet. My mother, my brothers, and I followed; Suzette threw her winter coat over her pajamas and hurried after us. We left the room and walked across a hallway filled with smoke, muddied with the foam and water firefighters were using to extinguish the blaze, and rode the elevator down to the lobby.

It was 17 degrees outside, with a wind that tore through our clothes.

My father took us to a coffee shop at the corner for hot chocolate and coffee, where we waited and wondered. Had we lost another home? Would we have to move again to a new hotel? Upstairs were all of our worldly belongings. It seemed unthinkable that the little we still owned could be destroyed.

After several hours we were told we could return to our rooms, which miraculously had suffered little damage. I was in an oddly chipper mood. I'd been completely vindicated in my habit of going to sleep with my clothes on.

In my mind, it was prudent to be on guard in this country.

AT LAST, OUR MEANDERINGS through Brooklyn paid off, and we found an apartment: four rooms, including the kitchen. It was far smaller than Malaka Nazli, hardly enough to accommodate six people, but at least it

was ours, and after nearly a year of hotel rooms, it seemed almost pala-
tial. It was on a street where families were either from Italy or the Levant.

Our elderly landlord, Basil Cohen—no relation to the dentist—
traced his ancestry to Aleppo exactly as my father did. After negotia-
tions worthy of two Syrian bazaar merchants, my dad and Mr. Cohen
agreed to a rent of $95. It was over our budget, but we felt under so
much pressure to move, we had no choice. Mrs. Kirschner wanted us
out of the Broadway Central immediately. She had threatened to stop
paying our bills, which would have effectively left us homeless.

We would be like normal people again, with a real address. It was,
as far as I was concerned, our most exciting day in America: we were
going shopping for furniture. We would have our own beds, chairs,
couches, tables—all that we'd missed for so long.

As we trooped to Macy's in single file, I noticed the cold didn't
bother me a bit. I spotted the sign from blocks away: "Macy's: The
World's Largest Store."

I was in awe. But once upstairs, as we wandered through the vast
showrooms, we realized there was nothing we could actually afford.

The salesman showed us magnificent king-size beds that looked as if
they were out of a movie set but weren't even remotely within our budget. Noting our dismay, he escorted us to a corner where Macy's kept
its least expensive merchandise. He pointed out several spartan metal
cots. The low-lying folding beds were small and forbidding, with thin
striped foam mattresses barely a couple of inches thick.

"*C'est comme dans l'armée,*" my mom remarked acidly; It's like the
army.

We walked out of Macy's having spent our entire furniture budget
on six folding steel cots.

Mrs. Kirschner blanched at the bill—$254—and accused my father
of being a spendthrift. Why Macy's? she demanded to know. Why not a
neighborhood shop?

She continued to see him as the cause of all our mishaps. A feminist
before the flowering of the feminist movement, she viewed my father
with such suspicion and hostility that even his attributes in her eyes
turned into flaws. Why did a refugee from Egypt shop only in first-rate

department stores? Why did he speak with an upper-crust British accent? she wondered. Surely it was an affectation.

My father had lived his entire life by a code of honor. In Egypt, he had been respected and admired precisely for his principles. Yet the chasm was so immense between him and our social worker she found almost nothing to admire—not even his lovely English. His insistence on tradition made him obdurate in her eyes. His devotion to faith and ritual was hopelessly quaint. She cast a wary eye on the religious passion that had always defined my father; because she was so secular, the product of a secular society, she didn't share that passion and dismissed it as superficial and devoid of sincerity.

There was also the notion that he was unemployable—or at least, that was the verdict rendered by *la Nyana* within weeks of our arrival. The agency simply couldn't envision a place for my father in the vast and abundant land of opportunity known as America.

"I have always worked, madame," he told Mrs. Kirschner. Though he had always been secretive with us about his business dealings, he spoke at length with her about his experience as a grocer, an investor, and a pharmaceutical and chemical salesman.

He was desperate to work. When Mrs. Kirschner pointed out his physical limitations, he exclaimed, "Le bon Dieu est grand." But this only led her to complain in her case notes about Dad's tendency to always invoke God. My father, she wrote, "resorts to denials, distortions, and evasion, and his philosophy is that 'God is Great,' which he constantly expresses in French."

She cast a cold eye on his impassioned plea that he *needed* to work to support all six of us, as he had always done. The social worker suggested he apply for welfare, instead. It was, again, a quintessentially American idea, certainly for the early 1960s. But nothing she said could have offended him more. He didn't want charity, he told her coldly. Besides, he had a better idea.

In his walks around Manhattan, he'd noticed the hundreds of little stalls and stands that were everywhere, in the subway stations, on street corners, by bus stops, near any crowded venue, manned by one or two people selling cigarettes, newspapers, chocolate bars, candy, chips, cookies, magazines. Now *there* was a business that seemed manageable.

It reminded him of the old days when he and Oncle Raphael had peddled groceries together.

He was prepared to start small, and besides, in his mind, these micro-businesses had enormous potential: New Yorkers wanted their morning paper and their Almond Joy and their pack of Camels in the same way that in Cairo, the typical Egyptian could be counted on to purchase a bottle of olive oil and a can of sardines.

My father decided he was going to open a candy store.

He started combing the classifieds for newsstands and tobacco stalls that were for sale. If no one in America would hire him, it seemed the ideal solution. He decided to appeal to Mrs. Kirschner and *la Nyana* to help him. A loan of $2,000 would do the trick, and then he would be able to support my mom and the rest of us entirely on his own as he always had.

Mrs. Kirschner wouldn't hear of it.

She didn't think she was being arbitrary or unkind. On the contrary, she felt she was being solicitous of my father, whose limp had gotten worse in the months since we had arrived. Prominent doctors the agency consulted said he should stay off his feet and give himself time to heal, yet there he was, proposing a venture that would require him to stand all day. Besides, he didn't even have a coherent business plan— only supreme self-confidence that he could support us.

My dad's impossibly modest wish was turned down. The man who had done business with Coca-Cola couldn't be trusted to sell cigarettes and bubble gum.

In the middle of January, a major blizzard hit New York and left more than a foot of snow. It was more snow than we thought possible. A few days later, we left the Broadway Central for the second floor of the Cohens' brick two-family on Sixty-sixth Street in Brooklyn. Mr. and Mrs. Cohen were waiting to greet us. "Etfadalou," they cried, Arabic for welcome, and with typical Syrian hospitality, they offered us a platter of *khak*, salty ring-shaped biscuits covered with sesame. We hadn't eaten them since Cairo, and biting into the delicious treats made us realize both that we were far from home and that we'd finally arrived. The cots from Macy's were waiting for us. We still didn't have a table, and there was one chair for all six of us.

Yet even here we couldn't quite escape Sylvia Kirschner's wrath.

Six months after our move, she decided to make a home visit. That morning, my father asked me if I wanted to go into Manhattan with him. I nodded yes, eager to accompany him on what seemed like an adventure. I didn't realize that my dad was whisking me out of the house so I wouldn't run into our social worker.

The two were now openly at war, any semblance of civility gone. He had watched as she befriended Suzette, encouraging her to flout his authority by telling her that in America, it was fine for a young woman to be independent. My sister was now threatening to leave the family and live on her own. My distraught father called Mrs. Kirschner and complained she had sent Suzette hurtling down a path that could only lead to disaster. "We will be ruined, madame," he told the social worker. She shrugged and scribbled in her notes that he was being "extremely melodramatic."

My father had other plans for my sister.

At the end of our block, my father had found a new home for himself—the Congregation of Love and Friendship. There it was, the old Cairo synagogue he thought was lost forever, resurrected from the dead, even down to its original Hebrew name, Ahabah ve Ahavah. The Congregation was warm and inviting, and he was reunited with several of his old friends from Egypt, who had undertaken the same sad journey. They prayed with the familiar melodies of Cairo Jewry, in the cherished cadence and rhythm of the temples around Malaka Nazli.

Many of the men had sons Suzette's age who were eager to get married and rebuild their lives. He told the social worker he had suitors lined up for my sister. He couldn't help boasting how skilled he was at arranging marriages—he had helped each one of his five sisters find a husband. Surely he could make a fine match for his own daughter.

Mrs. Kirschner wasn't impressed. In America, girls didn't have to be married off while they were young. They could leave the hearth, pursue an education, have a career. She didn't think my sister had any obligation to get married—or to obey my father.

Dad found all of this unconscionable. On that hot summer day, he determined that he wasn't going to let Sylvia Kirschner get anywhere near me.

I helped him carry the large brown box he carted everywhere these days. My father hadn't found a job, but he was working. He had become a necktie salesman. Inside the box were dozens of ties, soft and silky and patterned in the most wonderful shapes and colors I had ever seen—a treasure trove that any adult male would be certain to want.

An hour or so later, Mrs. Kirschner arrived to find my mother alone. Where was my father? she asked. And where was I? She seemed dismayed we weren't all there, as she'd specified. She was also annoyed. What on earth was Leon doing taking a little girl out on such a scorching day?

My mom tried to soothe her. She brought out a platter piled high with cakes and cookies, and some lemonade, and said I had gone with him to work.

Sylvia Kirschner was beside herself. She decided that he must be using me to boost his chances of making more sales. With my dark hair and dark eyes, I "could easily attract attention," she scribbled furiously. She couldn't imagine why he would take me with him "unless of course, it was for the purpose of using" me to "get a sympathetic reaction" from customers.

I was very lucky: decades would pass before the country embraced a "Take Your Daughter to Work" day and little girls began joining their dads in cubicles and at computer screens and in corporate boardrooms, and having the time of their lives.

It was clear he was struggling in his new business venture, and there were days he didn't make any sales. But on that hot summer morning, as we walked hand in hand, he was hopeful and tender and solicitous. He smiled as he asked me, "Loulou, tu vas m'aider à vendre les cravates?"; Will you help me sell some ties?

He thought that I would bring him luck.

CHAPTER 17

The Hebrew Lesson

That first year in America, I often woke up with a start after dreaming of Pouspous. Lying there on my Macy's cot, I'd think about my cat in Egypt and burst out crying. Had Pouspous even survived? I'd wonder. Had she managed on her own, with none of us to look after her on Malaka Nazli? I was so agitated my father had to be summoned to reassure me, though I was past the age when I trusted him as completely as I had the day we left Cairo.

Each one of us fixated on an object or being that emblematized what we missed about our lost life. For me, it was Pouspous. For my mother, it was the feel of the heavy sapphire ring on her left index finger, how the stone glistened and caught the light. For César, it was a boyhood friend named Gaby, a young Coptic Christian who had idolized him and looked up to him; now, my brother carried Gaby's picture in his wallet. Isaac would later recall the particular angle the sun fell in the room facing the alleyway, the most wondrous room in the house because it had so many incarnations: as Zarifa's bedroom, Dad's office, and, at the end, storage room for our suitcases.

My sister, rebellious and defiant, insisted there was nothing she missed.

It was the reverse in my father's case; there was nothing he didn't miss about Egypt, though perhaps it was the roses he longed for most of all. His favorite complaint about our fall from grace between Cairo and Paris and New York concerned the flowers. They had no scent, he lamented.

To my father, the flowers of America were odorless and lifeless—artificial even when freshly gathered, and altogether inferior to the flowers we had left behind.

Leon was particularly upset about the roses. Lovingly cultivated by our Italian neighbors, they bloomed by the hundreds up and down Sixty-sixth Street, in the front yards of our working-class Brooklyn neighborhood, yet they emitted not a hint of perfume. Whether purchased from the corner florist or picked from a nearby bush, they still had no fragrance, a fact that filled him with a kind of existential despair, a sense of all that was wrong with our New World. How stark the contrast to the sprigs of jasmine whose perfume filled Cairo's night air, to the lilies and honeysuckle that grew wild in the streets, and to the roses, above all, the roses, the small, red, overpowering damask roses, descendants of the very first roses to grow on earth.

Even more unimaginable, stores sold plastic flowers, often for much less than what the real ones cost. They seemed extraordinarily alluring to me, and I kept clamoring for them, oblivious to my parents' angst.

Woolworth's, the five-and-dime located a couple of blocks away from us, featured shelf after shelf of artificial flowers that sold for only pennies a stem. There were also pretty, slightly more expensive silk flowers. Until we moved to New York, I had never seen artificial flowers, and I found them utterly exotic, a beguiling symbol of the new land.

I noticed fake plants in neighbors' homes and plastic centerpieces on coffee tables. Some families even kept them behind glass, on display along with the silverware and crystal, as if they, too, were valuable. These houses had shiny plastic slipcovers on their sofas and armchairs, and Formica tables in their kitchens. I found it wonderful.

I wanted a plastic slipcover, and a couch to go with it, a Formica table, and, above all, plastic flowers.

In a walk through Woolworth's, how I longed to gather up the imitation tulips and daisies, to pluck them from their green Styrofoam moorings and fashion a bouquet to decorate our new apartment, which was stark and barely furnished and badly in need of cheer.

My mother firmly said no. "Loulou, ça suffit, non c'est non," she snapped; "Loulou, enough already, no means no."

My father, who tried to indulge me in so many of my requests, also made it clear he wouldn't contribute a penny.

My father could never acknowledge to us how much he longed for the texture of the life left behind. He fixated instead on the flowers as an emblem of all that was bewildering about his new home and his new country, and all that he missed about his old home and his old country.

Perhaps that is why we still hadn't bothered to unpack. The twenty-six suitcases were securely stored in the basement of our first American apartment. Many still contained exactly what had been placed in them two years earlier. My mother never retrieved the pretty polka-dot dress *la couturière* had made for her on the eve of leaving. My father's brocade robe remained neatly folded, where we had packed it.

Nobody wore brocade dressing gowns in Brooklyn. None of the clothes the tailors and dressmakers had sewn for us those final weeks seemed appropriate, somehow.

Dad didn't stay out late anymore, yet he was still a night creature, and rarely fell asleep before dawn. Wearing cotton pajamas and *sheb-shebs*—vinyl house slippers from the five-and-dime—he simply buried himself in his old prayer books. They were among the few items we did retrieve from the suitcases.

Even as we were beginning to feel settled, Suzette declared that she was leaving. Although she had been threatening to move out for months, none of us had taken her seriously, least of all my father. His reaction alternated between fury and desperation. What so many American families would view as part of the natural order—a daughter growing up, longing for independence and a place of her own—was anathema to him. He regarded the prospect of her departure as the worst misfortune to befall the family since leaving Egypt.

His rages and invectives proved to be too much for Sylvia Kirschner.

She was relieved of our case, no doubt voluntarily. My father blamed her personally for Suzette's rebellion. Dad was sure our family would be disgraced, that news his unmarried daughter had left home would spread among the ragtag group of refugees. Honor, reputation, social standing—that is what really mattered to him, far more than money.

Our new social worker, Shulamit Halkin, also cast a cold eye on Dad's old-world concerns and what she, too, saw as his patriarchal tendencies. But she didn't seem quite as beguiled by my sister as her predecessor had been. Suzette, bored with her clerical work at the First National City Bank, had confided to Mrs. Halkin her desire to become a doctor and save the starving children of India, and for that matter of the whole world.

"She could not quite explain to me what motivated her interest in all the children of the world and what she hoped to do for them," Mrs. Halkin noted wryly in the case files. It was hard for her to see why my sister worried about the sufferings of distant peoples in Asia when her family was barely making ends meet in Brooklyn.

There were times my father couldn't even afford the $95 rent, and Basil Cohen had to look the other way. The tie business wasn't as lucrative as he'd hoped. He wasn't making nearly enough to pay the rent, and if Suzette left and stopped chipping in, our financial situation would go from precarious to dire.

Newly enrolled at the local elementary school, I was shielded from most financial worries, as was my brother Isaac, who attended junior high.

César was carrying all five of us on his frail eighteen-year-old shoulders. That first year, he went from one menial job to another, never staying more than a couple of weeks or a couple of months: salesman in a Syrian-owned record store, clerk at a French bank, until finally settling down in an entry-level post at an American conglomerate named Continental Grain that was blessed with a cosmopolitan, amiable culture.

I shared a bedroom with Suzette, and assumed we were ideal roommates. I didn't realize that I was a seven-year-old nuisance who cramped her twenty-year-old's style in a thousand different ways. While she yearned to flee, to come to terms with her emerging womanhood, I wanted to hunker down and reclaim the childhood I had lost. While she kept sparring with my dad, I felt closer to him than ever in his alienation and distress. While she thought of our house as impossibly confining and wanted only to run away, I loved its simplicity, the fact that we were no longer on the run.

My formula for achieving permanency and stability? I would decorate our room on Sixty-sixth Street with white voile curtains. I had glimpsed them fluttering about the display windows of home furnishing stores up and down Eighteenth Avenue, our shopping mecca, and they became my version of the all-American fantasy of the white picket fence.

"What about the white curtains?" I asked the night before she left home for good. I was watching her fold clothes into her new suitcase, not one of the twenty-six. I couldn't quite manage a light tone. I felt wounded and every bit as betrayed as my dad by her departure.

She shrugged. I would have my white curtains one day, she assured me. "But maybe not in this room on Sixty-sixth Street," she added. That was how I knew that she was really leaving, how I grasped it before my parents, who still thought this was all a big bluff by their contrarian, wayward daughter.

My father wasn't about to use my brand of gentle suasion.

"Mogrema"—Criminal—he shouted as Suzette left.

The door slammed, and there was silence. A few minutes later, I heard my father's halting steps down the stairs, and the door closed again.

I LIKED TO STARE out the curtainless window of the room I now had all to myself. Up and down the street and in the adjoining blocks were other families like mine, new arrivals from Cairo and Alexandria. In the mornings, I'd watch my father walking to attend the first service of the day at the Congregation of Love and Friendship. His step seemed painfully labored.

Sometimes, we wouldn't see him the rest of the day. He'd stay for the second set of prayers, intended to draw those laggards who couldn't make it by 6:00 a.m. Afterward, he'd linger, most likely because he had nowhere else to go, but also perhaps because he didn't want to come home. Deeply upset about Suzette's departure, powerless to stop her—or persuade Mrs. Kirschner to stop her—he withdrew to the little shul down the street.

He sat by himself, somewhat removed from the other men, though he enjoyed amicable relations with all of them. They'd banter, laugh, trade rumors, discuss how their lives were going in this confounding new land. The men relished exchanging the latest gossip every bit as much as their wives and daughters, who came only on Saturdays and sat crammed together behind a tall concrete wall, chatting nonstop in the makeshift women's section.

My father, on the other hand, sat there silently, and spoke up only when a reader, or even the rabbi, had made a mistake—and then it was to call out the proper wording or intonation. He had such a prodigious knowledge of the liturgy, he could recite by heart almost every single prayer. He was meticulous and so strict he didn't tolerate the slightest error or deviation in the reading of a sacred text.

He was a biblical security guard—a policeman of the holy word—challenging anyone who dared to change what was perfect and deathless and immutable. When it came to questions of God and religion, my father was once again the Captain, a figure of authority. Some of the older congregants were annoyed by his interruptions. But Dad was unyielding, and forced them to reread the offending word or phrase.

Oddly, the younger boys didn't mind; they deferred to him on all religious matters, and dubbed my father the tzaddik—the saint, the holy man.

When the synagogue emptied out, as the men left to go to work or home to their wives, my dad would linger. Sometimes, Rabbi Halfon would join him. The ancient little rabbi, who had reigned over the original Congregation of Love and Friendship, hadn't seen my father since Egypt. Their reunion on Sixty-sixth Street had been one of the unexpected joys of moving to America. The two liked to sit together at the same table, reading. They didn't exchange a single social pleasantry, yet they were the closest of friends, joined together by their shared passion for the holy texts.

Even after the rabbi went home, Dad stayed. If the door was open,

I'd catch a glimpse of him as I walked to school, a lonely, solitary figure bent over a large book. His lips moved silently, and he didn't even see me when I'd wave. His injured leg stretched out, his cane perched on the side of the table, he fidgeted and shuffled from side to side to alleviate the pain. But then he'd usually settle into a position he could tolerate, sit back, and resume reading.

He would peruse the weekly portion of the Bible, review sayings of the prophets, study obscure codes of Jewish law. His favorite readings were the Psalms, the heartfelt pleas to God authored by King David. It was customary to recite them in a group, typically on the anniversary of the passing of a loved one, but my father liked to read them by himself every day. He'd go through the psalms one by one, and read the entire book from beginning to end.

He sat in the same spot from the late morning through the afternoon and early evening, without a break, helping himself to little snacks left on the table by the caretaker—colossal black olives and pita bread. The bread and olives were his lunch and dinner. He'd look up in the late afternoon and see a crowd of men entering. It was time for the evening services, and he hadn't returned home, and he hadn't sold any ties, and he hadn't made any money.

Elie Mosseri, twenty-two and newly married, liked to sit near him when he could. He was from our old Cairo neighborhood, and he and his family had lived, like us, in a building on Malaka Nazli. He had vivid childhood memories of Leon in Egypt, a tall, commanding figure who always sat by the synagogue's altar, and didn't hesitate to speak up when a word in a prayer was misspoken.

He was shocked at the change he saw in him in America. Elie plunked himself into a chair next to my father and began to read from the large frayed Hebrew books that came from the library of the abandoned Congregation of Love and Friendship in Cairo.

Elie could tell Dad was struggling, that he didn't have a job or very much to do. Why else would a man stay nine, ten hours in synagogue?

The congregation was booming. New immigrants descended on it day and night. They prayed in the exact way they had in Egypt, determined to allow nothing to change, despite the fact that they now lived thousands of miles from Cairo.

———

AN ENTERPRISING PAIR OF brothers began baking pita bread and delivering it to homes up and down Bensonhurst. Morris and Joshua Setton discovered there was an eager clientele willing to spend what little they had to avoid eating white bread. They later opened a grocery store of their own that stocked only Middle Eastern delicacies, realizing early on that merely eating the hot round loaves of pita was a crucial step in retrieving our lost life.

That was the goal now: even more than forging ahead in America, we wanted to replicate what we had left behind in Egypt.

Grown children were encouraged to marry other Levantine Jews— not Americans, not even other American Jews. Mothers taught their daughters their favorite recipes so they would be preserved and passed on, and their children and grandchildren would dine on the same foods generations of families had enjoyed, a cuisine that mingled the best of Syrian and Egyptian traditions, the sweet fruity passion of Aleppo and the garlicky oniony zest of Cairo—stuffed grape leaves, stewed okra in lemon and tomato, pockets of lamb filled with rice and pistachios, meatballs in sour cherries.

Assimilation?

It wasn't a word we knew or cared to learn, and my father, finally reunited with his own people, was enjoying a modicum of peace and contentment. That was also why he stayed longer and longer at the Congregation of Love and Friendship, not deigning to come home.

It was also, of course, why my sister left. She now lived in a tower high up in the sky in Queens, a part of New York I'd never visited and that sounded distant and foreign. Her rebellion was nearly complete. She had moved out before getting married and against my father's wishes, and she had even sidestepped the unspoken rule about living in a ground-floor apartment, as had always been the tradition in our family starting with Malaka Nazli.

How foolish—and hopeless—to try to reproduce a vanished world, Suzette thought. It depressed her, this effort to build a Cairo-on-the-Hudson. There was nothing in common, she felt, between those humdrum streets of one- and two-family homes and itty-bitty shuls of Brooklyn and the energy and joy and magnetism and life that could be found in even the shabbiest Cairo alleyway. It was altogether pathetic, this attempt to recapture the past through groceries and synagogues.

Suzette remembered an exuberant culture where religion mattered, but so did going out at night and reveling in all the Levant offered. Our father, who now all but lived at shul, was the prime example of this dual existence, where faith and ritual had in no way hindered his ability to lead a rich and pleasure-filled life. In Egypt, it was easy to be religious and worldly at the same time, but that seemed an impossibility here in America.

It was as if you arrived and were ordered to choose one door or the other, not both.

The biggest missing ingredient was glamour. It had been the defining quality of Farouk's Cairo in the 1930s and '40s—the elegant British officers, their beautiful mistresses, the passion for dancing, the insistence on going out all night every night. Even after the revolution, there was still an allure to life that couldn't be obliterated by even the most heavy-handed and ruthless military rule. Restaurants continued to stay open late, and people still didn't dine till midnight, and then they went out dancing or merely sat back to watch the belly dancers.

New York, supposedly the greatest city on earth, seemed to be limited to ten or twenty blocks that were drab, functional, and utterly devoid of style.

The moment we arrived, Suzette had sought out a dear friend from Egypt, Marcelle, who was also in Brooklyn. Marcelle had been a wild child, very brainy, very chic, the archetypal party girl. But somewhere on the way to New York, she had undergone a transformation. Shortly after settling here, Marcelle became engaged to a very devout man, and affected the habits and garb of religious American women. She changed her clothes and even her name; instead of Marcelle, she called herself "Adeena." On the eve of the wedding, she was preparing to cut her hair and don a wig.

Suzette was stunned when she saw her irrepressible girlfriend wearing a long, modest dress and behaving deferentially toward her fiancé.

For Marcelle's wedding, my sister wore a sleeveless sheath. The moment she entered the synagogue, a group of women rushed over in a panic to drape her shoulders with a sweater or shawl. Under no circumstances, they said, could her arms be bare. That was when she'd sworn to herself that she would leave, and have nothing to do anymore with this community of expatriates who called themselves Egyptians but bore no resemblance whatsoever to the people she had known back in Egypt.

A couple of weeks after my sister moved out of my room on Sixty-sixth Street, my mother moved in. She took over the narrow steel cot left vacant by Suzette.

I didn't even question our sleeping arrangements. I took it for granted that my parents no longer shared a room; they hadn't even on Malaka Nazli. At first, it seemed necessary—we simply had no space. But as one by one my siblings left, and they continued to sleep apart, I realized that their refusal to occupy the same bedroom was part of a deeper mystery I could never hope to grasp as a little girl.

EACH MORNING, I LEFT hand in hand with my mom for PS 205, the small public school around the corner. She refused to let me go alone, despite the short distance, and I cringed because we attracted attention. Only when I had pushed the door open would she relinquish my hand and kiss me good-bye.

I was intensely aware of the differences between me and the other children. It wasn't simply my broken and heavily accented English. There was also the way I looked and dressed—like a Parisian *lycéene* who had landed in working-class Brooklyn.

My mom loved to dress me like the idealized Parisian schoolgirl of her imagination, sending me to school in navy blue sweaters and matching plaid pleated skirts, though of course, back in Paris, confronted with real French schoolgirls, I dressed like an Egyptian.

I marched out of the house looking as if I were on my way to some swank school in the sixteenth arrondissement. Alas, my wardrobe only set me apart once again from my classmates, the little girls who came in colorful dresses with wide flare skirts—clothes hand-sewn by their mothers, elaborate and showy, and lined with ribbons and bits of velvet trim.

"They are so *fellahi*," my mom exclaimed when I expressed a longing to look like my schoolmates. To compare someone to an Arab

Loulou as a young schoolgirl in Brooklyn.

peasant was her most searing put-down. To Mom, these families of seamstresses and housepainters and sanitation men were merely fella-hin, although I noticed when I went to their homes that they lived far more comfortably than we did.

I wasn't sure how we were superior to them. We were the ones who were poor and struggling desperately; wouldn't that make us fellahin?

Lunch was also an ordeal. I noticed that most of the children carried lunchboxes—small, sturdy, efficient affairs, made of metal. Inside, there'd be a sandwich consisting of two slices of white bread filled with tuna or bologna or ham or cheese. My fare came in a brown paper bag, typically a fresh roll purchased at dawn by Mom from the Italian bakery around the corner and stuffed with a Hershey bar.

My classmates peered at my lunch and then at me. "Is that a choco-late sandwich?" they asked, their eyes widening.

I nodded yes and took a big bite. I found it absolutely delicious— that is, until I realized they found it awfully strange, this *pain au*

Loulou's American class.

chocolat lunch of mine. Self-conscious, I asked her not to make them anymore.

I wandered around my new school, shy as could be in my pleated skirt and sweater sets. Classes, at least, were simple. I was more than prepared for the rudimentary arithmetic, reading, and science that constituted an American elementary school education. My problem was gym. I had never played sports in Cairo or Paris, so the sudden emphasis on physical fitness took me completely aback. I wandered through gym classes in a fog, volunteering for nothing.

Without explaining why, Mom instructed me never to reveal I came from Egypt. I told anyone who asked that I was from Paris. To my teachers, this made me a charming novelty, "that little French girl," and even my closest friends didn't know the truth. Yet I often felt I was living a lie, burdened with a dark secret that was bound to surface: that I wasn't French at all, that I was born in Cairo.

AFTER SUZETTE LEFT HOME, I noticed that my brothers also seemed to drift away. César, our sole means of support since my sister had moved out, worked during the day and then went off to school and after that to his mysterious nightlife, so that I was asleep by the time he came home. Isaac spoke of joining the U.S. Air Force. I was secretly pleased at the prospect of his leaving. He and I were always clashing, and unlike my oldest brother, who tended to be protective, I found Isaac, six years my senior, caustic and combative.

While César was more docile, and closer to my dad, even he seemed increasingly removed from the religious practices that had been so central to our life. My father's worst fears about America were coming to pass. He'd order my brothers to pray. They'd merely shrug and go their own way.

Even my mother was rebelling: outfitted with a new pair of dentures, Sylvia Kirschner's last legacy, she kept hinting that she wanted to go out and find a job. Her first declaration of independence came when she decided to abandon the Congregation of Love and Friendship. She'd tried valiantly to fit in, but the area where women sat, a cramped space behind a wall, was so dismal, and it was often impossible to find an empty seat or follow the service. Even worse were the busybodies who kept grilling Mom on Suzette's whereabouts. "Is she married yet?" they'd ask, while the less kindly among them posed the same question with more edge: "She *isn't* married yet?"

My mother would manage a valiant smile and blithely insist that her daughter, only twenty years old, was too busy going to college and minding her studies. None of the women believed her.

And so we fled a congregation where we had found neither love nor friendship. My mom took me by the hand one Saturday morning, and we walked past Dad's synagogue and didn't stop until we arrived at another house of worship called the Shield of Young David. It was slightly more diverse; while worshippers were also from the Levant, there were Turks, Moroccans, Tunisians, Algerians, even Mexicans, in addition to the Egyptian and Syrian Jews.

It was also roomier, with an airy women's section whose graceful filigreed wooden partition made it possible to follow the entire service

Edith in the backyard at Sixty-sixth Street, Brooklyn, our first real home outside of Egypt.

without straining. Taking her place in the first row, with me at her side, my mom sat down and never looked back. She began to make friends with the women around her. Her closest friend, a Moroccan immigrant she called Madame Marie, had a daughter, Celia, who was slightly older than me. Mom and Madame Marie confided to each other their angst about raising daughters in America.

I followed Mom's lead. My best friends were Diana, a Syrian girl, and two sisters Grace and Rebecca whose parents were Turkish and Mexican. I looked up to Diana's older sister, a pretty, sedate teenager named Marlene, with dark hair and dark eyes, who bore an aching resemblance to my own sister. I felt at home behind the partition, and at ease with these girls in a way that I never did at school.

My mother never set foot in Dad's synagogue again. The two prayed apart as they ate apart and slept apart and lived lives apart, except when it came to worrying about "*pauvre* Loulou," and then they behaved like the most intimate married couple, so that I felt as if my maladies and crises were what kept us together as a family.

I BECAME PART OF a bold experiment in our expatriate community: I was going to Hebrew school. The ancient language had long been an exclusively male domain, and the move marked a dramatic break with the past, an embrace of modernity and America's egalitarian ways. Historically, women like my mom would sit in the synagogues of Cairo and Aleppo listening to prayers they could neither read nor comprehend.

Hebrew school was every night of the week except Friday and Saturday, including classes early Sunday morning. The primary goal was still to educate the boys, of course, as they would have their coming-of-age or bar mitzvah ceremonies within a few years. The boys sat in the front of the room, enjoying the lion's share of attention from the rabbis. Girls meekly took their place in the back, several rows behind.

It was more cynicism than enlightenment that prodded our elders to educate us in the same way they were teaching our brothers. America was such a seductive society, and from their perch in Brooklyn, the rabbis recognized they faced a formidable threat—greater even than the anti-Semitic outbursts the community had occasionally faced in the Middle East. Their strategy was to circle the wagons—pushing faith and religious observance for *both* sexes as the antidote to a secular life.

My teacher was a kindly, avuncular rabbi whose name, Baruch Ben Haim, meant the Blessed Son of Life. He had a no-nonsense approach to teaching. Unlike most of his rabbinical colleagues, he tended to treat boys and girls equally and gave us almost—almost—the same amount of attention.

He had devised a fearsome system to gauge our progress. At any given moment, he could call on us and ask us to begin reading from a random text. We were supposed to keep reading until we made a mistake; then he'd count the number of words and verses we had read without stumbling. The stars were the children who could read a sentence or two or more; the dunces stumbled after a couple of words.

It was all terrifying. I worried that he'd assign a text I wouldn't be able to read, and I'd become the laughingstock of my class. I turned to my old Arabic teacher for help, confiding in Dad my struggles with Hebrew. He listened quietly, then handed me his prized red prayer book and signaled to me to read.

The Hebrew lesson had begun.

I read a few words, haltingly at first, but when I saw he wasn't stopping me, I gained more confidence and continued. If I made an error or had trouble pronouncing a word, he would interrupt and show me the proper way to say it. He did so mildly, with the infinite patience he had shown when, at his bedside in Cairo with Pouspous in tow, he had taught me to speak Arabic by having me feed the cat pieces of cheese.

Every evening, when I came home from Hebrew school, I'd join my father for a lesson. He would choose a book at random, open it at a page that suited his fancy, and point to a line he wanted me to read. Sometimes he selected a hymn, and he and I would chant it out loud together, with me hastening to follow him. Dad had a surprisingly strong voice and could carry a tune, so he was able to teach me melodies he had learned as a child and had carried inside him for more than sixty years.

With the constant practice, the black letters began to make sense. I felt completely at ease leafing through the ancient books that were lying around the house. I realized that I had a natural affinity for Hebrew—a facility that I didn't enjoy with Arabic or English.

My father had become so much more reclusive since coming to America, and I, at seven going on eight, had devised a way to penetrate his hard shell. It wasn't by confronting him head-on, like Suzette, or defying him, as my brothers were doing. It wasn't even by engaging in idle chitchat like the men at the Congregation of Love and Friendship.

Leon with his red prayer book, Brooklyn, 1965.

Rather, we grew closer when I became his prayer companion, like Rabbi Halfon and Elie Mosseri. By sitting next to him and sharing his passion for the words that danced and floated on the pages of these impossibly frayed books, I formed a bond with him that transcended both the words and the pages.

The lessons paid off in ways neither of us could have predicted. At Hebrew school, the rabbi would call on me, expecting me to falter after a couple of minutes, only to find that I could now read page after page, prayer after prayer, fluently, without tripping, and could decipher even the most complicated passages.

Dad chuckled when I told him of my progress. Then he pointed to a new text and asked me to read: the lessons had to continue exactly as before.

I was a far more passionate disciple than my brothers. I would grow up chatty and extroverted and expansive, able to make friends in a heartbeat and conduct animated conversations with the most diverse figures, relating to them on the basis of a mutually shared interest in food or literature or movies, but I would never feel as close to anyone as I did to my father those evenings we sat quietly side by side, mouthing phrases in a mysterious language, from books whose yellowed pages crumbled in our hand as we turned them.

CHAPTER 18

The Ballad of
the Tie Salesman

In the summer, when I was home from school, I loved to accompany my father to work. It was a treat for me, since most of my friends were away at camp or on vacation with their families, and I was left to my own devices. Mom was also relieved: she knew how restless I was in hot, deserted Bensonhurst, and was at a loss as to how to keep me busy and entertained when she barely had money for the household. "*Pauvre* Loulou," she would sigh, what I needed were *des vacances,* she'd declare. But her idea of the perfect vacation—an upscale sleep-away camp in the Swiss Alps—like so many of her notions for me, was fanciful and utterly unattainable.

I was riveted by what my father did, so setting out with him to Manhattan was like going on holiday. It was hardly a conventional job, being a tie salesman. But it was the only work he could get—requiring little capital and no infrastructure of any kind save a large cardboard box in which to cart his merchandise.

Dad had no boss and no regular hours, he drew no salary, and his "office" was the streets and subways of New York. That is where he would buttonhole prospective customers, opening the brown rectangular box that was overflowing with dazzling ties of every hue.

They all bore elegant labels that said "100% Silk" and "Made in France" or "Made in Italy." Strangers would pause, taken by his charm and the beauty of some of his merchandise. And perhaps they were also moved by the sight of this tall, dignified old man trying to earn a living in the streets and train stations of New York City.

I'd see him heading for the door with the flat box that was the centerpiece of his business. He dressed neatly, donning even on the hottest days a jacket and tie, along with a jaunty straw hat. In addition to the box, he would carry a large yellow manila envelope that served as his briefcase. It contained small swatches of fabric that were central to his newest line of work.

Unable to earn enough as a tie salesman, Dad was now trying to branch out into textiles, a field that he knew well from Egypt. I wasn't exactly sure what he did with those pretty squares of silk and cotton that he stuffed in the manila envelope, though I always eyed them longingly, imagining how lovely they'd look on my dolls. I was always hopeful he would offer me one or two pieces of fabric.

Dad would signal to me to join him, and together we'd walk, ever so slowly, to the subway station on Twentieth Avenue, two blocks or so from our house. It was after the morning rush hour, so the station was almost deserted, only a few stragglers here and there making the commute to Manhattan. As we waited for the train, he'd try to spot prospective customers. They were mostly men he would approach, and he reached out to them so politely and deferentially, using only the mildest of sales pitches.

"Monsieur," he'd say, lightly tapping them on the shoulder. With a rustle of suspense, he would lift the lid off the brown box to reveal dozens and dozens of ties, in an impressive array of prints and colors. There were bold ties, made of satiny red or blue fabric, and ties with classical Ivy League stripes, ties with whimsical paisley prints and ties that were sober and severe.

"These cravats are one hundred percent silk," he assured a potential customer, deliberately using the French word for tie. "They are all imported from Paris and Rome."

With a smile, he offered to make them a good price—a discount—if they purchased more than one. Occasionally, a sale was consummated then and there, on a bench at the Twentieth Avenue station in Benson-hurst, as we awaited the N train to Manhattan. More often than not, though, the stranger looked but didn't buy, and my father graciously put the ties back in the box, closed it shut, and, without revealing a hint of impatience or weariness, tucked it back under his arm. Together we boarded the subway, choosing two seats together, and settled in for the long ride into Manhattan.

Eager to make myself useful, I eyed the passengers sitting near us, eager to spot potential customers. I'd whisper and point to a well-dressed gentleman looking idly our way, or to a couple who seemed friendly and were smiling. Some welcomed the diversion, and willingly took a look at Dad's treasure trove. But other riders were cold and wary or self-absorbed, and refused even to acknowledge us.

The first stop was Canal Street in Lower Manhattan, a nerve center of the textile industry that was filled with fabric stores.

One morning that my mother had gratefully turned me over to Dad's care, relieved to have a day to herself, I noticed that my father was looking me up and down. He was clearly displeased. He disapproved of the dress I was wearing that morning, a simple cotton frock fished out of a bargain-store bin. It had cost only a couple of dollars, but I liked it because it was light and jaunty and, to my eyes, a becoming shade of yellow. Discount stores flourished all around our neighborhood, and they were a favorite destination for my mom, who was grateful for places where she could actually afford to shop.

My father was visibly perturbed. "Loulou, est-ce que c'est ta seule robe?" he asked; Is this your only dress? He knew that I had other clothes. What he meant was: Why had I worn such a cheap, poorly made outfit to accompany him? I knew the drill. From the time I'd started going with him to the Nile Hilton, when I was little more than a toddler, he had taught me to always look my best when it came to work and seeing *des clients*.

Still, I felt sheepish and confused, taken aback by his anger. My mother was always complaining about money, how little she had to spend on me, and she encouraged me to hunt for bargains. Yet there was my father, clearly embarrassed by my appearance.

Though he was an old man in an inexpensive seersucker jacket and straw hat, in his mind he was still the boulevardier he once had been, when money was no object and he could afford to dress himself and his young daughter in the finest Cairo had to offer.

By the time we stepped off the train, my father was no longer scouting customers to buy ties; he was focusing on his second line of work. The area around Canal Street was lined with textile stores. Although the main garment district was uptown, the stores here did a brisk business in "remnants"—cheaper, discounted fabrics that were often leftovers.

I was still feeling self-conscious in my yellow dress when we entered our first shop. I managed a smile, determined to be on my best eight-and-a-half-year-old behavior, and listened carefully as my father made his sales pitch. He pulled out a swatch of shiny brocades from the bulging yellow envelope and handed them over to the owner.

"Monsieur, I can get these for you at two dollars a yard," he assured him.

"How much will you want?" the owner asked as he fingered the small squares of fabric glistening with strands of gold and silver thread.

My father acted nonchalant, as if he were prepared to do the merchant a favor with the sale. He wanted almost nothing for himself, he suggested—at most a minimal fee to oversee the transaction and as compensation for his troubles. "A couple of pennies, monsieur," my father said with a smile. "Two or two and a half cents a foot."

How can you divide a penny in half? I wondered. But I forced myself to stay silent.

The owner kept studying the brocade samples, holding them up to the light. He nodded hesitantly, unwilling to commit himself. "Let me think about it," he finally told my father, who took the fabric squares back and placed them back in the envelope.

We walked out in the July heat, and ambled down Canal Street to a store a couple of doors away. The routine began all over again, with another swatch. Then we journeyed to another, and then another after that. There were stores that specialized in cottons, and others that sold fine lace to be used for curtains or doilies or even wedding dresses, and one establishment that featured bolts of fake fur, only fur. We walked in, and my father reached into his magic yellow envelope and pulled out samples of leopard prints and black-and-white zebra stripes.

I prayed that this particular deal wouldn't go through, not out of disloyalty to Dad, but because I longed to use the samples to make fur coats for my dolls.

At last, we reached a vast store that seemed more upscale than its neighbors, whose sign said it specialized in silk. The owner greeted my father warmly, all the while peering curiously at me. "This is Loulou," my father said by way of introduction, "my granddaughter."

The owner, more amiable than the other men I'd encountered so far, softened and motioned to us to step inside his office. On display were a raft of pictures of his children and grandchildren. "Loulou, your grandfather works so hard lugging his samples from store to store in this heat," he said to me.

I nodded, unsure what to say. I was thoroughly bewildered. Had my father made a mistake in English? Didn't he mean to say that I was his *daughter*?

My father proceeded to retrieve some samples from the seemingly bottomless manila envelope. They were incredibly soft to the touch, in the vivid colors that were all the rage in the summer of 1965—hot pink, lemon yellow, lime green, electric orange. As I weighed my options, trying to decide whether or not to speak up, a deal was consummated— the first of our long, hot day. We walked out hand in hand, with the owner enjoining me to take good care of "Grandpa."

"Loulou, you should tell your grandfather to retire, a man his age is too old to be walking around on a day like this," the owner added. I nodded again, not daring to say a word.

My father, clearly buoyed, announced it was time for us to dine. He loved food, and unlike Mom, who was so self-denying she would eat nothing more than the crust of the slice of pizza she would buy me, Leon had held on to some of his more freewheeling ways. He was fond of café fare, and tried to find it in a city of coffee shops and diners that served mediocre dishes in charmless surroundings.

How far away he felt from La Parisiana or the other elegant brasseries with their vast terraces overlooking the Nile that had been his favorite haunts. They were illuminated by strings of colorful lanterns, and as the sun set, the lanterns began to emit a soft glow that was reflected in the water, so that the river seemed to be a thousand colors at once. Sipping a mug of ice-cold beer while staring at the Nile, Leon would at last order dinner. It was always *poisson grillé*, his favorite dish, and he could sit there for hours, staring at the lights and nibbling the delicately grilled whole whitefish.

We searched for a coffee shop where we could rest. We never took taxis in New York, even on days like this when the sun was beating down on us, and the temperature crossed the 90-degree mark, and the pavements around Canal Street seemed to melt under our feet, and my father's step became much more labored. We simply walked and walked, until we reached a large cafeteria, delighted to be able to enter an air-conditioned palace with cushy booths.

My father made it clear that because of my hard work, I could order whatever I wanted from the menu. I settled on a chocolate egg cream, tall and chilled and bubbly from the spritz of soda water. He ordered a treat for himself as well, a strawberry milk shake. He had a passion for strawberry milk shakes, which struck me as a very wonderful dish for an old man to love—all pink and white, ice cream and froth. Because I decided his beverage was tastier than mine, because I always wanted what he had, he kept passing me his tall pink glass and letting me take sip after sip from a straw.

The stop had a magically restorative effect. When we left, we had enough energy to make the rounds of several more fabric stores in the grueling sun.

I never thought to ask my dad if he was anxious about approaching businessmen he didn't know. He rarely complained, so it didn't occur to me to wonder whether, in this city whose culture and manners were new to him, he ever felt intimidated, or scared, or upset or humiliated when he was rebuffed, which was far too often.

My father was always intensely private about his disappointments. For example, once NYANA turned down his request for a loan to open his own candy stand, he never spoke of it again. Similarly, I had no idea how much he was struggling making a living through selling ties, and now, bolts of fabrics.

I did ask whether he had ever held down a "real" job.

"Jamais de la vie," he replied; Never. He thoughtfully amended his answer, recalling how as a very young man, he had briefly held an entry-level job with a Cairo bank.

Why didn't he stay at the bank? I asked.

"I didn't like having a boss," he said.

He was trying to rebuild, now, and forever ensure his independence. He had even gone back to dabbling in his true passion—*la bourse.* My father was obsessed with the stock market. It was his one remaining vice, as he had given up every one of the other pleasures he would indulge in back in Cairo. There were no more poker games, no casinos where he could gamble the night away with a king or a pasha.

But the New York Stock Exchange, even he would concede, was every bit as thrilling as *la bourse* had been in Egypt. When the market was up, he was up. When it was down, he fell into a deep funk. You would think he had millions of dollars tied up in his investments, but as far as I could tell, there were only some small holdings made possible by the little bits of money he earned selling ties, or the occasional $50 or $60 cut he obtained on his half-penny commissions brokering textile deals.

He favored only the most exotic investments—Zambian copper mines, which was traded on the Lusaka Stock Exchange, for instance, or Consolidated Gold Fields, a South African mining conglomerate founded by Sir Cecil Rhodes, the man who established the Rhodes scholarship and had dreamed of British domination of Africa.

My dad also bought shares in an energy company that would drill for oil in Angola, Equatorial Guinea, the Russian Republic of Tatarstan, and, most tantalizing of all, Egypt: our family may have been forced to leave, but Dad seemed still to root for his old homeland. Another favorite was the defense contractor Sperry-Rand. The Vietnam War was heating up, and Leon quickly realized that a maker of military equipment and technology would be a sound investment.

His heart—and mine—was with the Zambian copper mines. When I asked him about the strange-sounding stock, he winked and made it sound as if that single investment held so much promise that any day now we would find ourselves the owners of our very own African copper mine.

My mother didn't share his fascination. "Ton père jette son argent," she'd say loud enough for him to hear; Your father throws away his money.

Dad loved Dow Jones and the industrial average. But as much as he reveled in the rock and roll of the market, and believed in the implicit promise of *la bourse,* as he insisted on calling it, Edith loathed the very idea of investing, and any time she talked about stocks and bonds, it was with contempt. The stock market was gambling, she believed, pure and simple. She saw it as a dangerous indulgence by a grown man living in hardscrabble Brooklyn, who still thought of himself as a carefree Cairo boulevardier.

My father had never stopped loving games of chance. Even when he'd join us at the beach, carting slowly his green-and-white beach chair until he found us, he barely dipped his sore leg in the ocean. Instead, seated in his beach chair, a cold soda in his hand, he continued to pore over the pages of the *Wall Street Journal* or the stock listings of the *New York Times,* anxious to see how his investments were faring, as diligent a monitor of the market as any captain of industry or Wall Street swell.

THERE WAS ONE MORE stop to make. Together, we boarded the subway for the short ride from Canal Street to Delancey Street, in the heart of the Lower East Side. We didn't have far to walk, but by the late afternoon, my father's limp had become more pronounced, and we kept stopping so he could rest. I asked if I could help carry the brown box, but he refused. I was never allowed to lug either the box or the precious manila envelope.

Years before it became part of the downtown revival, the area near Delancey Street had a more animated feel to it than the textile district. It was thronged, and the stores seemed to offer a thousand different wares, which they exhibited outside their shops on the sidewalks—from men's hats and fine women's clothing to pots and pan and other housewares. Many of the stores were run by men in skullcaps, Orthodox Jews, many of them Holocaust survivors and their children, who had brought the skills of the old Eastern European shtetl to the Lower East Side.

Leon at the beach, Brooklyn, 1960s.

I followed Leon to a dingy back street of tenements, where the shops were so small, they didn't even have signs to identify them. One especially careworn storefront required us to walk down a few steps and push open a large metal door. It was a factory, where I saw women hunched over sewing machines in a cramped, windowless room. They looked up briefly as we walked in, then went back to their stitching. My father made a beeline for the owner.

The two seemed to know each other well. The factory made ties, and my father had come to restock his supply. The owner brought out several samples from his newest line and spread them out on a small counter. They all seemed beautiful to me, and I nodded eagerly as my father indicated his preference for this style or that. Finally, Dad asked the owner: Were there labels on the ties?

The owner sheepishly admitted he hadn't gotten round to that.

My father was visibly upset. "Monsieur, I'll have to come back another day," he said, and started to leave.

246 • LUCETTE LAGNADO

That is when the owner signaled to an older woman who sat at the head of a large rectangular table, where at least half a dozen seamstresses were ensconced at their machines, and gave her rapid-fire orders in a language I didn't understand. She nodded. He turned to my father and assured him that the work would get done if we waited a bit.

That is when I understood. The ties that I had admired for so long hadn't been imported from Paris or Rome. They were made here, in this small dusty sweatshop somewhere between Essex and Delancey, then had phony labels proclaiming their exotic provenance attached by one of these women bent over their Singer sewing machines.

And they weren't even made of silk.

We walked out with a new brown cardboard box stuffed with a fresh supply of ties that bore the elegant but deceitful labels. My father seemed so buoyed by his brand-new stock that he cornered a group of men on the street and was able to make a sale.

As we walked toward the subway with our inventory, I avoided asking about the "imported" ties. Instead, I pressed my father about what I should do with my own life. At nine, I was restless and ambitious beyond my years. I wanted to be a detective, a secret agent, a master spy. I wanted to travel around the world and return to all the cities we had fled. I wanted to hold an important job and make large sums of money and take taxis constantly.

"You can have a little job," Leon said in that mild tone that masked a definite intensity.

We both knew that it was common for women to work in America. My own mother was still flirting with the idea of employment. She had received one lucrative job offer from Grolier, the prestigious French publishing house.

She had actually dragged me to the interview, much to the surprise of the senior editor who had asked Mom to come in after receiving her charming and mellifluous application letter, in which she touted her passion for books and her longing to work in the literary world. My mother hadn't wanted to leave me home alone, but hiring a babysitter would have been unthinkable—I never once was left with one. And so she took me along.

I'd sat silently, afraid to breathe as the Grolier editor frowned and made some comment to the effect that she hoped I wasn't a noisy child. Mom had been able to reassure her that I would keep still, and went on to impress her with her unusual credentials—a knowledge of literature that went back to her Cairo childhood, followed by a stint as a teacher and librarian at a prestigious boys' school called the École Cattaoui Pasha. She even told of the Key—the Key to the Pasha's Library.

An offer came days later. My mother was stunned. She had never believed Grolier would actually want to hire her, let alone pay her so bountifully—beyond our wildest imaginings. The job came with a title, entrée into the fashionable Manhattan neighborhood where Grolier was situated, and a salary that seemed to us almost absurdly high. It happened to be greater than what my older brother and father together were earning.

Most important of all, she would be working once again with books.

Still, after a great deal of agonizing and soul-searching, Mom turned Grolier down. It was in part her own inability to break loose, to free herself from her role as a wife and housewife. Then there was her fear of defying my father. She knew, of course, that he didn't approve of women working: hadn't she quit a job she loved, working with the pasha's wife, when she married him?

It was too ingrained a part of Egyptian—and Syrian—culture: a woman, especially a married woman, doesn't work. We were no longer in Egypt, and my father couldn't support any of us anymore, but that was almost beside the point.

Yet she was clearly ambivalent. There was a piece of her that loved books even more than marriage and a family. She also knew how perilous our finances had become since Suzette's departure. Now, suddenly, she was being handed a chance to reverse our shaky fortunes. There was no question that the position at Grolier would have propelled her—and us—into the middle class and, if she had done well, perhaps the upper middle class, an almost unheard-of achievement for a refugee family. But she was heeding an unspoken rule about women and work, no matter how antiquated it felt in America, where women were entering the workplace by the millions and society was changing to accommodate them.

Only my father didn't change. He cast a dubious eye on all these developments. Whoever heard of a woman working? he chuckled. It was part of his inviolate code of conduct that women—certainly a wife or a daughter of his—should steer clear of the rough-and-tumble business world.

I asked him, if I had to work, what should I do?

"You can have a little job," he said finally. The "little job" he had in mind took me aback. He suggested that I open a flower shop.

"But you don't even like American flowers," I pointed out. His idea seemed out of left field. I had no interest in plants or flowers. He didn't seem to hear me.

"Loulou, you can open your own flower shop," he repeated, and there was a faraway smile on his face, as if he were smelling the roses of Egypt. But his final piece of advice on that summer day was sparse and to the point: I had to marry, and marry well—*un banquier,* a banker.

I had to find a man who would give me back all we had lost when we left Cairo.

CHAPTER 19

Waiting for Elijah

In the weeks leading up to Passover, the holiday when it is forbidden to eat bread, my mother reenacted a ritual with me that dated back generations. She enlisted my help in the ceremony known as the Sifting of the Rice, in which each and every grain of rice had to be pristine, cleansed of all its impurities.

Mom was in the throes of a massive spring cleaning, mopping and scrubbing every corner of our apartment. Even a small crumb, left behind by mistake, would be a desecration. It also risked unleashing my father's outrage. I followed her around, mother's little helper, lugging buckets filled with hot soapy water that she needed to wipe the walls and windows. Sometimes, when she wasn't looking, I joyfully flung the water onto the panes of glass and the Formica countertops, letting the suds drip all over the floor. I would take a cloth and, seeking to emulate her, would scrub scrub scrub scrub with a passion and abandon I never brought to housework.

Not that I was ever expected to do much of it.

It was an unspoken commandment at home: "Loulou shall do no

menial chores." I was the envy of my friends, who were constantly being called upon by their moms to wash dishes and dry dishes, to set the table and clear the table. I wasn't asked to do so much as rinse a single glass, or sweep, or dust, or make my own bed, or anyone else's.

This was largely my father's influence, his unwavering conception of how a daughter of his should be raised. In Cairo, where there had been an endless supply of maids to attend to every aspect of running a house, a pampered daughter was never expected to help.

There were no maids in America, or none that we could afford, but my father still objected to seeing me do chores that had been the purview of *les domestiques*.

He thought nothing, however, of having my mother perform all the household tasks. Such was the hierarchy in an Oriental household, be it Muslim or Jewish: the father as patriarch came first; his rule was absolute. The sons came next. They wielded almost as much authority, and were treated like young princelings. The daughters commanded their own sphere of influence by virtue of their closeness to the father.

The mother came dead last; her role was to cater to everyone else.

Still, Mom didn't quarrel with my dad about what I should and shouldn't do. She agreed that I should be spared mundane tasks, and encouraged rather than resented my privileged status. My upbringing was one of the few areas in which my parents were seemingly in perfect accord. But their motives were starkly different. My father expected me to follow a traditional path of marriage and a family, but was confident that any man he'd embrace as a son-in-law would provide me with all the help I needed. He didn't discourage my studies at school, figuring they could enhance my desirability as a mate. But he saw no need for me to learn to cook and clean.

From time to time, he alluded to the dowry he was amassing for me, suggesting I would one day have my pick of husbands.

My mother had a different rationale for encouraging my sloth. She wanted to make sure I had *none* of the skills that would help me cope as a wife and mother. In a sense, she was the quintessential passive-aggressive. Cowed by my father, she also had a profound rebellious streak, which showed itself in the way she was raising me. She shuddered to think that I would gravitate to a life of domesticity, that I, too, would become a prisoner of Malaka Nazli (or its American equivalent).

God forbid, she would tell me, that I would grow up and have to cater to some man and an ungrateful brood of children.

For good measure, she'd add: "Never marry a Syrian." She suggested that Syrian men were handsome and seductive, but also bossy and narcissistic. They treated women like slaves, she told me. I didn't dare challenge her, and even at a young age, I found myself looking at Syrian men warily.

When I developed a crush on my friend Diana's older brother Maurice—Syrian, to be sure, but with wonderful blue-green eyes—I worried lest I be forever a captive of his charms.

Mom's strategy to discourage me from the life most of my friends dreamed of was simple and diabolical—she refused to teach me any of the basics. I wasn't shown how to clean a room or make a bed. I didn't learn how to fix lunch or dinner. Of course, that also meant I was never shown how to prepare the wonderful dishes she had mastered during those rough early years with Zarifa. Mom had emerged as a wondrous cook. Night after night, she prepared foods that were redolent of the Levant, and that I would never taste again because I'd never been shown how to make them—tender pockets of lamb stuffed with rice and pistachios and cooked for hours in the stove until the meat seemed almost to melt off the bone; okra flavored with so much lemon and garlic, the entire kitchen was filled with the scent.

I was left to my own devices. I slept late, read novels, and languished about the house while she scurried about from errand to errand, chore to chore.

The exception was Passover, when the workload was so overwhelming she couldn't cope on her own. I didn't mind pitching in because even the most ordinary household tasks seemed imbued with a higher purpose. Besides, my mother seemed panic-stricken, though I wasn't sure if it was a fear of God or a fear of my father that motivated her.

Every Passover, Edith took me by the hand to the basement, where the twenty-six suitcases from Egypt remained, neatly stacked. In what had become a ritual, we opened one after the other in search of the ancient metal box I viewed as the source of all the mysteries and plea-sures of the holiday.

Round like a hatbox, it was made of a burnished dark gray steel that had weathered all our voyages before finally ending up in our apart-ment in Brooklyn: from Cairo to Alexandria, Alexandria to Athens, Ath-ens to Genoa, Genoa to Naples, Naples to Marseilles, Marseilles to Paris, Paris to Cherbourg, Cherbourg to Manhattan, and, at last, Man-hattan to Brooklyn.

The box lay undisturbed deep inside the brown leather suitcase that had been its first home, protected by Mom's astrakhan fur coat, her wedding gown, and other layers of ancient clothing.

I half expected ghosts to come flying out of the box, a djinn in a great puff of smoke. Instead, I only saw the same lovely porcelain dishes, wrapped in individual paper, we had put away the year before. There were china cups, small liqueur glasses, and sets of silverware—real sil-ver, not the 25-cents-a-fork variety we bought from Woolworth's.

Many of the items were almost a century old. They had belonged to my grandmothers Zarifa and Alexandra, and Mom treated them as if they were holy relics, fingering them ever so gently. There were tea-spoons so minuscule, they would fit inside a dollhouse. The spoons, including Zarifa's small spoon made of gold, had a unique function: they were used to sample the *haroseth*, the maroon-colored jam made from crushed dates and raisins that was the centerpiece of the Passover meal.

Though I was useful in helping Mom pick through suitcases and dust off silverware, nowhere was I more needed than handling rice. Back in Cairo, when rice came in twenty-kilo sacks replete with stray bits of grain or straw, it was important to sanitize it for the holiday. This was considered such a crucial task that the typical Jewish housewife took it upon herself to inspect the rice no fewer than seven times, each more rigorously than the last. Nor could the work be entrusted to the maids. Even in the relative comfort of Egypt, where almost every chore could be turned over to a hireling, women such as my mother per-formed this task themselves, or else with other family members.

Admittedly, some of the more progressive families went over the rice only four times. Edith insisted on seven. In Cairo, Mom had "hired"

Suzette and César to help her out, installing them at the large dining room table and paying them a few piasters for the work.

But here in America, when our rice supply consisted of the tightly wrapped cardboard packages of Carolina or the even pricier Uncle Ben's, we had on our hands a product that had already been processed, purified, homogenized, sterilized, and hermetically sealed.

What impurities could those milky white granules contain? What could possibly be sinful about eating Uncle Ben's?

Edith warned me to be careful as she spread a large white sheet on the dining room table, put me in a chair by her side, then forced me to dump out the contents of box after box so that the table was covered with mountains of rice. Each of us began the task of going through each grain of rice, separating the pure white grains from the blemished brown ones. I would put rice into a large bowl and proceed to review a handful of rice at a time, relishing how the grains felt as they slipped through my fingers.

I never questioned the necessity of what we were doing. It never occurred to me to wonder why we needed to peer at thousands upon thousands of individual grains of rice, one by one, one by one. I never challenged any of the rituals we followed at home, they were such an essential aspect of who we were as a family.

And among the scores of observances and sacraments that we tried to uphold through our peregrinations, from our home in Cairo to our refugee hotel in Paris, the welfare hotel in New York, and an immigrant community in Brooklyn, the Sifting of the Rice was the one that we never compromised, because it was permeated with the most sacred aura.

My mother would smile as I handed her bowl after bowl of the purified rice. I had carefully examined each one precisely seven times to make sure there wasn't the slightest imperfection. We sifted through twenty pounds or more, for that was how much the family would consume within the first couple of nights of the holiday.

When I grew older, and learned that my American peers, the Jews whose ancestors hailed from Eastern Europe or Germany, didn't eat rice at all during Passover because they considered it taboo—almost as much of a sin as eating bread—I was completely taken aback.

What care we had taken to make sure the rice we ate would pass God's own inspection!

And then there were the friends and acquaintances who startled me in a different way, by eating whatever they wanted on the holiday—even bread. Whatever happened, I wondered, to worrying over the integrity of a single grain of rice?

The Passover marathon culminated in a ghostly candlelit inspection tour conducted by my father. Late at night on the eve of the holiday, Leon, clad in his pajamas, holding a tall white candle in one hand and a prayer book in the other, led a nocturnal procession. As he shuffled from room to room and from corner to corner, Mom and I followed anxiously behind him, holding our breath as he peered inside closets, opened kitchen cabinets, rifled through bedroom drawers, and bent down to examine the floor, checking for stray crumbs. He looked like one of those detectives from those 1940s film noirs going over a crime scene with a flashlight, searching intently for any evidence of wrongdoing.

He was as rigorous as these mythical old Hollywood detectives. While there was a piece of me that enjoyed the theatrical aspects of the ghostly inspection, I realized this was deadly serious business for my father. Mild-mannered in most of his dealings with us, he was intransigent, uncompromising, and almost tyrannical when it came to religion. There were no shortcuts to faith, my father believed, no rules that could be bent or broken.

My mother alternated between trying to please him and rebelling against his despotic ways.

"Fanatique," she would cry out.

But she usually retreated, either because she was cowed by him, or because she, too, became persuaded that God himself cared that our little Brooklyn apartment was spotless for the holiday, and that He personally wanted to make sure the steaming platters of rice that would accompany the fragrant meat stews, or served alongside the thousand exotic delicacies she prepared, adhered to the highest standards of heaven and earth.

OF ALL THE ARCANE rituals and ceremonies I associated with the holiday, none was more sacred to me than shopping for a new dress.

My mom spent months carefully putting money aside so I could have a proper outfit to wear that single week of the year. For a holiday that signified renewal, where closets were emptied and shelves were stripped bare and floors were scrubbed and old food was mercilessly cast aside, it would have been unthinkable not to have a sparkling new wardrobe.

No other occasion, not even my birthday, warranted as major an expenditure.

La Eighteen, as Mom liked to call Eighteenth Avenue, with its cheery array of children's stores catering to families with small budgets, was our choice destination. It was my mother's favorite shopping venue, and in the course of walks either alone or with me, she had befriended most of the storeowners and was able to converse in fluent Italian she'd mastered as a child with those who hailed from Naples or Calabria.

La signora francese, they called her, or simply *la signora,* and beckoned her to enter their establishments, though she could rarely afford to make a purchase.

As I looked forward to turning ten, I was no longer willing to defer to my mother in the selection of this all-important dress. In the past, she had wielded considerable influence, nudging me toward one or another outfit because of budgetary constraints or her own definite tastes. This time, I was determined to buy a dress entirely of my own choosing.

I longed for the years when my shopping expeditions were with my father, who had exerted almost no influence on what I wanted to buy. He'd simply stand aside, chatting with the salesgirls—only the pretty ones, of course—as I fluttered about a store, trying on this dress or that. He became involved only when it was time to pay the bill.

My dad no longer took me shopping: the responsibility was now left entirely to my mother. I found her rather a sorry substitute—opinionated where he had been open-minded, penny-pinching where he had been munificent. As if sensing that she was falling short, she announced that she had in mind a special destination for my all-important holiday purchase.

"Loulou, allons chez Milgor," she said jauntily, and off we went to the most upscale children's store we knew, the one whose shop windows filled us with longing, yet which seemed so intimidating that we were always afraid to actually go inside.

For our weekly Saturday walk, my mother and I always set out for Milgor, though we deliberately left it for last. We wanted time to pause and savor its elegant window displays, crammed with an abundance of overpriced fineries we could admire but never possess.

Its windows were like theater, a small outer-borough version of the great Manhattan department stores. Come Christmas, for instance, red velvet reigned supreme: Milgor's windows featured clothes that were either made entirely of crimson velvet or had a velvet trim. There were dresses with lush velvet skirts and skirts with soft velvet sweaters, gleaming red velvet coats with black velvet collars and severe black coats with discreet red velvet collars. Even the shoes featured small black velvet bows.

As the weather turned balmy, velvet gave way to frothy lace. Pint-sized mannequins appeared, clad in white dresses with veils and bejeweled crowns that made them look like the delicate child brides of my mother's generation, the child brides of the Levant. They were only confirmation dresses, popular in this enclave of Italian Catholics, but I thought of them as miniature bridal gowns, and fantasized how wonderful it would be to wear one.

My mother was beguiled. This was how she longed to dress me, in clothes that were stylish and refined—the kind that her children had worn once upon a time when she had money and leisure and could pick and choose. Staring at the windows of Milgor became emblematic of our new lives: recalling what we had once enjoyed, despairing at all we could no longer have.

In the spring of 1966, the windows looked like a pastel rainbow—dresses in pale peach, icy pink, pistachio green, and songbird yellow drifted across a make-believe sky, frivolous and flirty and oddly provocative with their high waists and puffy sleeves.

Few of the clothes had price tags, or if they did, they were so small as to be almost illegible. In this hardscrabble neighborhood where the women stayed home while their husbands toiled as cops and firefighters, and most establishments billboarded their low prices, Milgor's was the only store with enough of an affluent clientele to survive with this approach.

The dress I loved stood alone in a corner. It was pale pink, with a white bodice, pink buttons, and a small white collar. I pointed it out to my mother with glee. I had made my decision instantly and at a glance, the way even I knew that life's most important decisions can be made, the way that my father had spotted my mother seated at the café in Cairo.

We pushed open the formidable glass doors that seemed to say, "Stay away, you're not worthy." Though we'd ogled the displays dozens of times, we had never actually entered the store. What was striking was the silence, the all-encompassing stillness. That and the fact that there was no merchandise we could touch and feel and examine. We were used to sifting through crowded racks and foraging inside bins and competing with bargain-hunters for clothing that had been tossed in large piles.

At Milgor's, clothes were kept in glass cases that were shielded from prying hands by long wooden counters. If a customer needed help, a salesclerk would silently unlock a case and bring out the desired items one or two at a time, like jewels.

Having scrounged up the nerve to enter, I was determined to make a purchase. I told the one salesperson who approached us that I wanted the pink dress. Without saying a word, she went to a glass case and re- trieved the longed-for garment in my size. Up close, the dress looked and felt like cotton candy. Made of the softest cotton, it was more shocking pink than pastel, but the bodice, which had seemed white in the window, turned out to be a delicate blend of pink and white polka dots.

I rushed into the dressing room to slip it on. When I emerged, my mother was deep in conversation with the salesgirl. I had found the perfect dress, I interrupted them to say. I'd make good use of it, I vowed. I would wear it to the Seder dinner. I would wear it to see friends. I would wear it to greet Elijah.

My mother didn't seem quite as enraptured, and even the Milgor clerk cast a wary eye on my frothy pink number. Edith had coopted her, enlisted her as an ally—she had the knack of taking complete strangers and turning them into her friends. She wanted to see other options.

The Milgor saleslady dashed off to the back and returned holding a dress so new it was still in its plastic sheathing. It had arrived the previous day, so there'd been no chance to place it in the window, she said, and lifted the wrapping to reveal a striking turquoise dress. It had the fashionable empire waist of the season, but with a demure, old-fashioned twist: on the front were five large embroidered tulips in different colors, as if blooming in some fantasy garden. The dress was strangely elegant—a cut above even what Milgor's typically featured.

For my mother, it was love at first sight. This was the dress she wanted for me. And I hated it, hated its primness and its silly tulips and, above all, the fact that my mother had contrived to find me a child's dress when I'd hoped that I was all done with childhood, and longed for the wanton abandon and frivolity of the pink dress.

She implored me to try it on but winced at the steep price the salesgirl quoted. It was two dollars more than the pink dress, which already cost more than what she could afford. I noticed that familiar anxious look as I headed toward the private fitting room.

I walked out transformed, the refined little girl of Edith's dreams, in a powerful rival to the pink dress.

"Which one will you be taking?" the salesgirl asked Mom.

She hesitated. I noticed that she seemed nervous, eyeing the tulip dress, doing some mental calculations to see if there was a way we could afford it on the small, haphazard allowance from my father.

"Madam, will you be taking the tulip dress?" the salesgirl repeated gently, if insistently. "Madam?" She tried to shake my mother out of her reverie by placing both dresses, the pink and the blue, side by side on the wooden counter.

Sadly, regretfully, with a sense of such profound yearning that even the clerk appeared moved, my mother said we weren't going to be taking the dress with tulips. We would be purchasing the pink. I of course was overjoyed. The dress with the sea of flowers was my mother's ideal, not mine. It matched her vision of *la jeune fille rangée*—the sedate, well-mannered young girl she was trying to raise, albeit in a place where no one seemed sedate or especially well mannered.

The pink dress made me feel giddy and grown-up, as if I'd been granted a taste of womanhood. The magic that I'd always sensed in Milgor's windows was finally in my grasp. I felt almost as if I could float, the way the pastel mannequins seemed to do in the window display.

IT WAS HIGH TIME for Elijah to arrive: that is how I felt on the eve of Passover.

On top of sifting rice, combing the house by candlelight for forbidden bread crumbs, buying my pink dress, and helping my mom out with the general housecleaning chores, I had added one more sacred ritual: purchasing a wine goblet for the prophet Elijah.

It was my attempt to improve on a time-honored tradition, although in my case the custom had turned into a small obsession.

On both nights of the Seder, a cup of wine had to be placed on the table, set aside exclusively for Elijah. No one was permitted to drink from this cup or even touch it. It was intended as a gesture of pure hospitality. If Elijah were to stop by, he would find he had a place at the table.

It was a charming allegory, one of dozens of symbolic gestures in a holiday crammed with them: re-creating the Exodus from Egypt by carrying a make-believe bundle over our shoulders, munching the flat bread of affliction to reenact our hurried departure, acting out each of the ten plagues—blood, frogs, vermin, beasts, boils, hail, cattle disease, locusts, darkness, and the killing of the firstborn—until at last crossing the Red Sea to step into the Promised Land.

Except that in my mind, there was nothing figurative about this holiday. Our family had suffered under a modern-day pharaoh—Nasser—and our exodus had been hurried and filled with trepidation. And so, when I was told that Elijah came to every house, I believed with all that I held dear that the biblical prophet would walk through our front door on that night.

His arrival was so tangible and concrete that I prepared for it and found myself anxiously listening for his footsteps.

I took religion literally, possibly far more literally than even the rabbis had intended. When my mom, shunning ordinary candles, lit a floating wick in a glass filled with oil and made a prayer for our well-being, I believed the flame in the glass contained supernatural powers. I'd close my eyes and make a wish, certain it would be granted, in the same way that a couple of years later, when friends traveled to Jerusalem to visit the Wailing Wall, after Israel had reclaimed it in the 1967

War, I'd send along notes to God containing my most profound long-ings. I'd instruct them to place them in the deepest crack they could find in the wall: God read the scraps of paper left for Him every night. My father had taught me there were random moments on any given day when the heavens were open, and if I happened to say a prayer at one of those moments, my wish would be granted.

As I walked home from school, I'd look up to the skies trying to di-vine if the heavens were open, and make outsize requests. I prayed my sister would move back home, and to have my dad walk well again. I believed in shrines and holy men, in the power of psalms and incanta-tions, and, above all, in the possibility of miracles.

In the spring of 1966, I decided to throw myself into preparations for Elijah's imminent arrival.

"Loulou, *magnouna,*" my mother told me, using the Arabic word for a crazy person. She was mostly amused by my fastidious attempts to keep the faith, but also alert to any sign that I was turning out like my father: too religious, a fanatic. Even her gentle sniping couldn't get me off course.

Of all the mystics who floated through the Bible, Elijah was my fa-vorite. It was, after all, he alone among men who had been spared from dying because God loved him so completely. Elijah, blessed with eter-nal life, embarked on a series of good works. I pictured him wandering the globe, a kindly old man shuttling from city to city, from home to home, to perform his miraculous deeds.

Would he really stop at 2054 Sixty-sixth Street?

That was the question that consumed me. I thought I had an answer—a surefire way to lure Elijah to my house.

There were a series of rituals associated with welcoming him, in ad-dition to filling his goblet. In Cairo, we had prepared our own wine, either squeezing grapes by hand or boiling several pounds of raisins for hours and hours in large vats mixed with sugared water and lemon. The result was a thick syrupy mix we thinned out by adding more water until we had a beverage that was light and sweet—not exactly fermented wine, but delicious. There were always dozens of small yellow raisins stuck to my glass, which I loved to spoon out and eat. But here in Amer-ica, even Elijah was relegated to drinking sweet purple Manischewitz instead of my mom's delicately limpid and airy homemade brew.

Then there was the practice of unlocking the front door so Elijah could enter anytime he wished.

Most families, including my own, had once left their doors wide open the entire night. On Malaka Nazli, we sat in the dining room by the balcony, so Elijah would have, in effect, two means of entering. But with lurid crimes like the murder of Kitty Genovese fresh in our minds, and a growing sense of the perils of our new urban landscape, observing the custom in New York seemed foolhardy.

In my mind, the prospect of Elijah's visit called for much more than small, halfhearted gestures.

A couple of days before the holiday, my mom watched me curiously as I went through our collection of dime-store glasses and cups. I was turning them over, inspecting each and every one. It was part of my quest for the perfect Elijah's cup, I told her; I rejected them one by one, even the wineglass I'd purchased the prior year.

"Mais ça suffit alors," she snapped; Enough, already. She had no sympathy for this particular compulsion, and besides, she was feeling overwhelmed by the work.

I turned to my father, imploring him to give me the money that I needed to buy the prophet a brand-new wine cup. My dad didn't even look up from his prayer book. He reached into his pocket, pulled out a handful of coins, and then returned to his prayers. I grabbed the money and ran out the door, vowing that this would be the year when Elijah would arrive.

"Dieu est grand," he muttered as he continued reading from his little red prayer book.

Clutching my change tightly, I headed to Eighteenth Avenue. Many of the shops had signs on their windows proclaiming "50% Off" or "Final Sale," and some displayed their wares outside, in large cardboard boxes, with the price scrawled by hand—25 cents for a plate, 40 cents for a juice glass.

It was always possible to go to Woolworth's, of course, where bargains abounded, and the selection was generous. I could have picked up a new wineglass for pennies, but in my mind that wouldn't do. The purchase of Elijah's cup demanded special care and concentration.

I walked up and down the avenue, engaging in some classic comparison shopping. I'd hold the wineglasses up to the light, inspecting them as carefully as if I were choosing a Waterford or a Lalique. I ended up at a small hardware store on the corner. The decision was agonizing. There was such an abundance of cups, goblets, and glasses. There were tall flutes with impossibly slender stems, and minuscule cordial glasses, and an assortment of imitation-crystal wine goblets. I settled on a flute, but it cost every one of the coins my dad had given me.

I went home with my purchase in hand, but with a heavy heart. Would it really tempt Elijah? I ran over to my father to show him our new acquisition. He looked up from his prayer book, glanced at the cup, and nodded.

Without explicitly saying so, he seemed to suggest we had indeed boosted our chances of having Elijah join us and need only await his arrival.

ON THE SEDER NIGHT, I helped my mom set the table using silverware from the ancient steel box. I made sure we each had our own silver spoon, and poured the wine into the new cup myself.

My father came from the synagogue and went straightaway to the dining room table. He wanted to make sure every item used in the ritual dinner—a hard-boiled egg, to symbolize the cycle of life, a bowl of salty water, to remind us of the tears we'd shed in Egypt, a bowl of red jam to conjure the mortar we used as slaves—was in its proper place. He made sure Elijah's cup overflowed with red, sweet wine, and nodded his approval.

At last, having determined all was where it should be, he sat down and began to lead the Seder. I sat at his side, as always, his prayer companion, trying to keep up with him even when he read at such a fast clip it was hard for me to get the words out.

My mother and both my brothers were there, but Suzette, who had by now moved to Miami, almost never came. Year after year I waited for her, too, like Elijah.

On that night when we asked ourselves why it was different from all other nights, Dad didn't retreat into his impenetrable shell. Instead, he chanted with gusto. My favorite moment was when he took the silver spoon and clinked it against his wineglass to punctuate the text. It made such a lovely tinkling sound, almost like music, as if on that night Alexandra, the gifted piano player, had joined us and was accompanying our hymns.

Still, no matter how loudly we sang, our holiday had become not a celebration of the Exodus from Egypt but the inverse—a longing to return to the place we were supposedly glad to have left.

At midnight, when we were still only halfway through—my father insisted on reading every single passage of the service—I ran to the door.

I stepped out a few feet to the stoop, my eyes straining to see through the April darkness. I was hoping to catch a glimpse of Elijah as he made his rounds. I imagined him drifting up and down Sixty-sixth Street, having a sip of wine here, enjoying a bite there. I wanted to be the one to welcome him, to usher him inside our modest apartment.

Elijah, I had been told, was the good-luck angel, the one who could make wishes come true. He was said to be among us, at times, though only animals and people with the most finely honed instincts were aware of his presence. When a dog cried, I learned, it meant that he had seen the Angel of Death approaching, but when he smiled and barked joyfully, Elijah was on hand. The days before the holiday I peered closely at the German shepherds my Italian neighbors favored, wondering if they had spotted Elijah.

I was prepared with a laundry list of requests. I wanted to ask him for a closet overflowing with Milgor clothes and a sofa with a plastic cover, as opposed to the twin bed that doubled as a couch when guests came. I wanted my mother to worry less about money and my father to walk tall again and sell a thousand ties.

But the street was empty, devoid of passersby and the usual contingent of neighborhood children hanging out on stoops. I returned to the table.

The Seder was punctuated by an enormous meal; afterward, it was hard to summon the energy to continue praying. One by one, members of my family went to bed. Only my father lingered, oblivious to his dwindling audience, dutifully going through every leftover hymn. He clinked his glass with his little silver spoon and kept on chanting by himself. My mom urged me to go to bed, but I resisted.

I was waiting for Elijah, I said.

By 2 a.m., we were completely done. We had left Egypt three years earlier, and had left it all over again tonight.

I wanted to stay up and watch the prophet arrive, to see him at our table, sipping from the goblet I had chosen with such care. I pleaded to stay awake, but my father only smiled and urged me to go to sleep. Before going to bed, I looked out my second-floor window, toward the heavens and the roofs of the other two-family houses, hoping to catch a glimpse of the prophet in motion.

Come morning, I dashed into the dining room to check the wineglass. I peered at it closely, lifting it toward the light.

The cup was as I'd left it. There was no sign that Elijah had stopped by. It hadn't been touched, not even a drop.

CHAPTER 20

The Captain at War

On her first morning in America, Stella Ragusa woke up at dawn and ran to her bedroom window, which faced Sixty-sixth Street. She noticed a tall elderly man in a white skullcap walking slowly down the block. His shoulders were hunched, and he used a wooden cane to navigate. From her perch on the upper floor of a two-family house, she thought, he looks exactly like the pope. She wondered if John Paul VI happened to be in New York.

Dazed and slightly feverish from the jet lag and the excitement of finding herself in a new home in a new country, Stella decided that in some way she couldn't explain, the pope had made the same voyage that she and her family had undertaken, from Italy to Brooklyn.

Her family was sound asleep, or Stella would happily have roused them and had them join her at the window.

To eleven-year-old Stella, who had never seen a Jew, my father was the Holy Father, miraculously transported to our neighborhood.

Stella and I met a couple of days later, when she darted across the street to introduce herself, and we became instant friends.

She decided impulsively that she loved my father. She was persuaded he was a holy man even after both her parents and I had patiently explained why he covered his head exactly like the pope, and she had learned all about his faith and mine.

My bond with Stella and her charming Neapolitan family couldn't shield me from the growing feud between my family and our new Sicilian landlords.

Our former landlord, gentle Basil Cohen, tired of his widowhood and anxious to join the rest of the Syrian community on Ocean Parkway, had found himself a new bride and decided to sell the house; the buyers were our Italian next-door neighbors, the Valerios. Up and down our block, Syrian Jews like Mr. Cohen were putting their homes up for sale, and clannish Italian families were snapping them up as fast as they came on the market. Suddenly, we were among the only Jewish families left.

While Mr. Cohen couldn't guarantee we'd be able to keep our apartment, we were still hopeful. We had always enjoyed cordial relations with our neighbors, and I was fond of Mr. Valerio, who hauled garbage for New York City yet insisted on calling himself a "sanitation engineer." His daughter, JoJo, was a few years older than me, but she had been especially welcoming when we'd first arrived in January 1964, introducing me to the Beatles and showing me her "I Love Ringo" and "I Love Paul" buttons—my first lesson in American popular culture.

But it quickly became clear we had to move. The Valerios told us they wanted the entire house, our apartment included, for the benefit of their elderly relatives from Sicily. Nothing could dissuade them—not even my mother's attempts at charming them into changing their minds or my father's efforts to settle the dispute "a L'amiable."

We were being forced to leave our home in the same way we'd had to abandon Malaka Nazli: under duress.

Except that this time, Dad wasn't going along with it.

My father was in full battle regalia, prepared to fight this eviction as he hadn't fought the other. He was standing firm, making it clear that he wasn't going to be intimidated or harassed into leaving his home—or in this case, his four-room apartment. He was, after all, the Captain—determined not to let anyone push him around.

One by one, all the other Levantine Jews had left. Our favorite grocer, Khasky's, was looking to open a new shop close to Ocean Parkway, taking his savory black olives and zesty feta cheese with him. Mansoura, the baker, was already ensconced in a small storefront on Kings Highway, the shopping street in the shadow of Ocean Parkway, preparing his famous platters of Oriental pastries for the families who were flocking to the area, the way he had back in Cairo, when Farouk liked to stop by his café in Heliopolis.

By the dawn of the 1970s, even the Congregation of Love and Friendship was poised to move again. In the same way it had transposed itself from Cairo to New York two decades earlier, it was now planning to abandon its cozy two-family house and follow its members. The departures meant the death knell for Sixty-sixth Street. Bereft as we were of so much we had loved when we left Egypt, our day-to-day lives in America had come to depend on access to Khasky, the grocer, and his intoxicating products, which were imported from Cairo and Aleppo and Damascus and Beirut. It was as if life would lose its sweetness absent his little bottles of *maward*, the fragrant rose water our mothers tossed into all their cooking, and we couldn't get by even one day without reaching into his large wooden bins filled with a dozen varieties of olives or sampling his feta cheese.

The solution should have been simple: move.

We should have moved because our world was disappearing, and we risked being left behind again, with nobody left to help us survive the rigors of American life. We should have moved because, with the exception of Stella and her parents, we didn't fit in with anyone on our block. We should have moved because there was nothing to keep us there.

Because of the dispute, a street that had been a haven for us turned into a war zone. Neighbors we had liked suddenly turned harsh and abusive when the news spread that we didn't want to leave. The Italian families sided with our landlord, and we were outcasts.

When I left the house in the morning, Vincent Valerio and his in-laws glared at me, and if I was with my parents, they tried to intercept us or block our way. Our landlords demanded to know when we were planning to vacate the apartment. They were absolutely adamant: they wanted the house, and they wanted it now.

My father would pointedly ignore them, and if they made any untoward remarks, he'd wave his cane and threaten to call the police and have them arrested.

Mom resigned herself to the idea of moving. She began discreetly scouring the area for apartments. It was all done in a fit of despair, in defiance of my father. She had a job now and she was busy. She worked at the Brooklyn Public Library's stately main branch. She was only a clerk, but it was as if she had been handed the key to the Pasha's library once again. Anxious for us to feel settled, she narrowed her hunt to a few blocks around our house, eliminating Ocean Parkway as an option, though it should have been the first area to search.

Finding an apartment that met all of our specifications was a tall order. It had to be inexpensive. It had to be on the ground floor. It had to be near a subway, both for my father, who couldn't walk very far, and for César, so he could commute to work. Gas and utilities needed to be included in the rent. And, of course, it should be within walking distance of the synagogue, although it was unclear which synagogue: one by one, they were all shutting down.

Any apartment that failed to meet even one of these conditions—on the second floor rather than the first, or four blocks from the subway instead of two, where electric bills were covered but not gas—was instantly ruled out. Mom's search turned out to be fruitless.

My father was refusing to budge. Nobody would ever take his home away from him again.

I sensed the hostility each time I walked down the street. Girls I'd played with since childhood wouldn't even say hello anymore. Occasionally someone called out, "It is not your house, you know," and I'd shoot back in whatever Brooklyn-girl tough talk I could muster, "It is too my house."

Yet Stella and I grew closer in the crisis. When she wangled invitations to a party her cousins were throwing, I eagerly—if nervously—accepted. My friend promised Italian food, Italian dancing, live Italian music, and, best of all, Italian boys.

I knew her cousins, who lived at the end of the block, only in passing. They also were new to this country and hadn't yet assumed the harsh demeanor that seemed almost a prerequisite of life in our working-class Italian-American neighborhood. They were outsiders, too, and I liked them instinctively.

Since my friend was intent on wearing her seductive new flamingo pink dress with the zipper down the front, I took steps not to appear hopelessly prim. On a sales rack at a nearby boutique, I spotted a short red cotton dress with a lace-up neckline that could be tied or untied to be chaste or daring. Belted at the hips, it fit perfectly, and though I could scarcely afford it, I bought it anyway—stealthily, without asking my mother for money lest she grill me about a party I was certain she wouldn't want me to attend.

Come dusk, I slipped out of the house, wearing my new red dress. "I'm going to meet Stella," I cried out as I left. My friend came down to meet me, resplendent in her pink dress, and we happily agreed we both looked fetching as we marched toward her cousin's house.

The party was nothing like the staid, all-girl affairs I was used to attending. In a cramped, wood-paneled basement, I could see clusters of handsome boys—or were they men?—in their late teens and even older, with smooth dark hair, dark eyes, and slender waists. Their tight, fitted shirts were left open almost to their belts, and they were smoking. They eyed us up and down as we entered. A small band was playing Italian music.

Stella's cousin came forward to introduce us. Almost everyone in the room was a recent arrival, and they seemed both more approachable and more vulnerable than the boys I knew. Most could barely speak a word of English. I found them intensely exotic, and decided I'd happily dance with any one of them. To my delight, they instantly dubbed me "L'Americana," and that became my name for the evening.

It was the first time I had ever been mistaken for a "native"—the dreaded, exalted American girl—and I was beside myself with joy.

It seemed extraordinary to think that in a crowded basement in Brooklyn, I had finally arrived.

"Che carina, questa Americana," a boy exclaimed as he walked toward me. He had ash-brown silky hair that fell over his eyes, an amiable smile, and a grown-up, confident manner. His arm around my shoulder, he led me confidently to the dance floor. Suddenly panic-stricken, I tried to catch Stella's eye, but she only nodded and smiled.

My partner couldn't speak a word of English, but that didn't seem to matter, not on this night, when the music was deliciously foreign, and I wore a daring red dress, and found myself close-dancing with someone who had movie-star good looks and could well have been in his twenties. I didn't know his name and he didn't know mine, and that didn't matter either. The only moniker I cared about on this particular night was "L'Americana."

And perhaps one other.

"What does the word *carina* mean?" I asked Stella during a break, when my dance partner left me to get a beer. "What does it mean *exactly*?"

Stella frowned. "It is like sweet, cute," she said, throwing out possible definitions.

I was thrilled at the compliment. I understood perfectly well that for the first time in my life, a man had found me pleasing.

I didn't leave my partner the entire evening. The louder and more frenetic the music became, the more other couples crowded around us, the more closely we danced. As we swayed to songs whose lyrics I didn't understand but whose tunes were enthralling, he tightened his grip, and I felt at last grown-up.

We left the party only when Stella, who didn't have as much luck finding a partner, insisted we had to go home or risk catching hell from our parents. It was well past midnight—the latest that I'd ever stayed out without telling my family where I was going.

"Ciao, carina," my partner told me, and he leaned over and kissed me tenderly. Sometime in the course of the evening, I'd realized that I was far too young for him. He had grasped that long before I did, yet continued to dance with me. He hadn't so much as left my side—either out of gallantry, friendship, affection, or desire.

Upstairs, both my parents were waiting up for me, but it was my father who seemed especially agitated. "Où est-ce que tu étais?" he demanded to know. I replied as nonchalantly as I could that I'd gone to a party with Stella and her cousins, and had forgotten the time.

"Il y avait des garçons là-bas?" he asked, getting straight to the point; Were there boys there? He was absolutely livid.

"It was very crowded, and there were lots of people," I answered evasively and in English. It seemed somehow the safer language.

Loulou as an American teenager.

"Tu vas ruiner ta réputation, tu vas gâcher ta vie," he shouted; You will destroy your reputation. You are going to ruin your life. He employed the tone and words he had used only with my sister.

For Leon, watching the social transformation under way across America in the late 1960s and early 1970s was like observing the goings-on on a distant planet, not his own—and certainly not the one he wanted his youngest daughter to inhabit.

Now that I was catching the full brunt of his anger, I felt more resentful than contrite. What was this reputation I had to guard so zealously? Honor, standing, community—the notions that consumed my Levantine dad—seemed so quaint and irrelevant in my world. What did the values of Old Aleppo have to do with life in New York?

OUR LIVING SITUATION HAD to come to a head, of course, and it did one morning when I was least prepared for it.

As I left the house with my father, we were stopped by Mr. Valerio and his in-laws. They were leaning against the stoop, their arms folded across their chest. My father continued walking but they moved to block his way. Mr. Valerio, standing alongside his mother-in-law, a heavyset woman who always wore black, began telling my father we had to leave at once.

"We are calling the marshals," Mr. Valerio declared.

His in-laws joined the fray. I had assumed they didn't know English, but the old woman, whom I dubbed the Black Spider, knew a phrase or two.

"Dirty Jews," she muttered. "Dirty Jews," she repeated more loudly, in case we hadn't heard her.

My father didn't even blink. "I am calling the police," he yelled back. "They will put you in jail."

I stood there petrified. At any moment, I was sure blows would be exchanged and then what? Who would defend my father—and me, for that matter—from these raging strangers I had thought were my neighbors and even my friends?

That is when I started to cry. As I stood a few feet from my house, facing an old woman hurling worse epithets our way than any we'd heard in Egypt, I suddenly found that I couldn't stop crying. Nobody seemed to take note of my anguish except my father. He was, momentarily, more annoyed with me than with our antagonists.

"Loulou, ne pleures pas. Il ne faut pas pleurer en face des étrangers," he told me; Don't cry. Never cry in front of strangers. Fight back, he hissed. As he fought the Battle of the Stoop, he was convinced I was only showing weakness, and weakness was the character flaw he could never tolerate.

The confrontation left me so shaken, I began to dread leaving the house lest I run into the Black Spider. Even more troubling was the prospect of a marshal coming to grab our belongings and turning us into refugees again. In my mind, it made sense to move immediately.

My father acted unperturbed. He wasn't afraid of a marshal, and he certainly wasn't afraid of an old woman.

The eviction notice arrived: we had to vacate immediately. Dad decided he would continue to fight, though his weapon of choice was hardly ideal. With no means to afford a private lawyer, he turned to Legal Aid. The organization that offered legal help to the poor was at the height of its powers, filled with cadres of idealistic young lawyers. My father—law-abiding, elderly, facing an eviction he didn't deserve—

could have been their poster child, and he was assigned a young fire-brand who instinctively hated landlords.

Our neighbors fought us in the quintessentially American way—by hiring an expensive private attorney.

The day of the hearing, we arrived in court looking like a Mediter-ranean version of the Hatfields and the McCoys. My neighbor was flanked by his wife and his mother-in-law. The Black Spider was dressed, if possible, more somberly than usual, as if she had stepped out of some Sicilian village. Their attorney, polished and elegant, stood at their side. He seemed seasoned in the ways of Court Street, the nerve center of the legal community, where judges, attorneys, and hangers-on made deals and reached understandings before even walking into the court-room. I noticed that he bantered with everyone from the judge to the clerk.

My parents and I also marched in together. I had skipped school on the advice of Legal Aid to show the judge "there is a child involved."

Our lawyer finally sauntered in. His hair was dirty blond and flowing, his beard bushy and unkempt, and for our legal showdown, he was dressed in corduroys and sandals. He came over to my father and put his arm around him. Why, he couldn't even imagine how we could lose.

The Valerios' lawyer, a small man in an expensive suit, began by making an eloquent plea before the judge. He argued that his clients—decent, hardworking immigrants—needed the entire house for the benefit of their frail relatives. As he made the case for evicting us, I heard my family described as thoroughly undesirable tenants.

While the attorney launched attack after attack on our character, our lawyer listened intently. Yet he never responded or stood up to yell, "Objection, Your Honor," as I'd watched Perry Mason do countless times on TV. He simply nodded and took notes as the other lawyer made the case that his clients were the true victims—not my refugee father, who sold ties on the subway to pay the rent, or my anxious mother, who wanted only to achieve a modicum of peace and content-ment in a country that, so far, had denied her both.

The judge rendered his decision then and there. Flashing a cordial smile at the landlord's lawyer, he declared that we had to move.

Our public defender seemed as stunned by the verdict as we were. He put his arm around my father once again and walked him out to the courtroom steps. My mother looked crushed. I felt that familiar long-ing to cry.

After the court hearing, my parents and I began searching in earnest for a new apartment in the neighborhood. I was terrified of being homeless. Would we have to move back into a hotel? What had become of the Broadway Central?

We grabbed the first place we found. It was a block away on Sixty-fifth Street, a breezy four-lane street my parents decided was like Mal-aka Nazli. Located on the ground floor of a two-family house, it was spacious enough for me to have my own room. My mother had wran-gled it for me by insisting to Dad, in one of her rare assertive moments, that I was growing up: "Loulou est une jeune fille maintenant"; Loulou is a young lady now.

Because it was at the front of the house, I could look out the window and enjoy the street life, my father told me wistfully, as if this were our house in Cairo. But at our landlady's insistence, our first act was to put up window shades. That is what people did in America: they installed screens and venetian blinds and drapes to prevent anyone from looking in, and to stop themselves from looking out. We'd never bothered with curtains on Malaka Nazli; only shutters that we hardly ever closed. But our new shades were thick and white and opaque, and I pulled them all the way down.

Our first Sunday in the new house, my father and I escaped by car service to Mansoura's. We sat down at the lone wooden table off to the side of the pastry shop while Isaac Mansoura himself went and fixed us a simple meal that wasn't on the menu. He cooked a pot of *ful me-dames*, the quintessential dish of Cairo: fava beans simmered in olive oil and lemon, with one hard-boiled egg floating in the middle.

We shared a large bowl, while the owner, whose friendship with my father dated back to Cairo, pulled up a chair and joined us. The two chatted in Arabic, and I hadn't the foggiest notion what they said, other than they looked supremely happy, my father dipping his bread into the flavorful brown sauce while Mr. Mansoura nodded intently, smiling at every bite my dad took of the dish he had so lovingly prepared. When we were done, he brought out a tray of assorted pastries, and only when we'd devoured the last piece of honeyed desserts did Mansoura call a car service to take us home.

We assumed that we could breathe easy, regain the peace we had lost. I missed Stella, of course: though we lived less than two blocks away, our friendship ceased after the move, and I never again saw the handsome Italian from the party.

But our new place had its advantages. Our landlords, the Cagnos, were a retired Jewish couple whose background was Eastern European.

We would never again be attacked merely for being Jewish.

SUDDENLY THERE WAS SO much space. Room after room, and almost no people and no furniture. We had brought the card table and folding chairs that had constituted our dining room set, and the six metal cots from Macy's, but we didn't own so much as a plant or a poster.

Our new landlords liked to come by and check on the apartment. They expected, no doubt, that we would become friends. We wanted only to be left alone.

A staircase led directly from their apartment to our kitchen, and they didn't hesitate to use it. We found it disconcerting to hear Mrs. Cagno's insistent knock. I noticed she was always peering at the other rooms, a disapproving look on her face.

Why didn't we have a couch? she'd ask. A dining room table? Drapes? Carpets? Chairs?

My father said nothing, and resumed praying. My mother was terrified of "La Cagno," as she dubbed her. Handling our landlady fell to me.

Left to my own devices, my instinct was to stall. I spoke of bedroom sets on the way, of a velvet sofa arriving any day. I almost believed my fanciful stories. But then the promised pieces failed to arrive, and our landlady grumbled again. I begged my parents to take me furniture shopping. They looked at me, bewildered. What was the point of buying couches and credenzas? It is not as if the six of us would ever live together as a family again.

Suzette and Isaac were gone, never to return. And now, even César had shocked us by starting to look for his own apartment.

One morning, as I walked slowly with my father toward Twentieth Avenue, where he was going to find himself a synagogue, I spotted Mrs.

Cagno making her way toward us, a squat figure with a loud voice swad-
dled in a dark shawl and a skirt that touched the floor.

"Why aren't you people taking better care of the apartment?" she
demanded. "Why are you people leaving it unfurnished?"

She was shouting, and she kept referring to me and my parents as
"you people."

My father kept silent, leaning on his cane that he didn't even raise,
then resumed walking.

I didn't. If the Captain had lost the will to fight, I had found in me a
well of rage and indignation. "You leave us alone," I cried. "You go to
hell."

She seemed taken aback by my fury, but only for a moment.

"You people would be better off in a tent in the desert," she said,
shaking her head before waddling away.

I stood with my father on the corner. Neither of us said a word; I was
trembling. We parted, and he resumed his daily quest for a temple. We
had reconciled since the night of the red dress and now were joined
together in our sadness. I had finally grasped the lesson from our earliest days in America—that I should tell no one that I came from Cairo.

It was so clear now: to be from Egypt meant you were from a primitive country, backward, unsavory.

For weeks and months, I replayed the scene with Mrs. Cagno. I
should have called her a liar, I thought. I should have told her my family had never lived in a tent, that we had owned a lovely apartment on
a broad tree-lined boulevard named after a queen, with a maid, a balcony, and a cat with a coat of many colors. I should have said I had attended a private girl's lycée that taught me more by the age of six than
I'd learned in all my years at my American elementary school. I should
have informed her that only an unfortunate twist of fate had placed us
at the mercy of fellahin like the Cagnos of Sixty-fifth Street. As for the
tall silent man at my side, I should have let her know that once upon a
time, he had bantered with princes and gambled with a king.

We didn't know any princes or kings anymore and my father had
long stopped playing poker and frequenting casinos. Within a few
weeks, we moved to another apartment down the block.

CHAPTER 21

The House of Prayer

W hen I least expected it, the Cat Scratch Fever seemed to return.

The symptoms were eerily the same as ten years earlier. A light, almost imperceptible fever that came and went. Night sweats. A sense of sluggishness that made it hard to keep up with my classes or take part in the bustle of activities that went with being a high school senior. And it was hard to fall asleep at night, no matter how tired I felt. I was only sixteen in the winter of 1973, but I felt exhausted, more like sixty, or how I imagined a sixty-year-old must feel. Most curious of all, the swelling at the top of my left thigh made a reappearance. There it was, the strange little bulge I had first noticed as a little girl in Egypt. It seemed a bit bigger, harder.

I tried not to look at it too closely.

The only good change, as far as I could tell, was that I was losing weight. Stepping on the scale each morning, I noticed that I was effortlessly shedding pounds. I could at last mimic the American girls I so admired with their slender waists and close-fitting jeans, who wandered airily through the corridors of New Utrecht, my Brooklyn high school. I would happily have traded all my Levantine curves for the svelte, compact figures of my classmates.

I didn't dare tell anyone at first how I was feeling. Surely, I thought, the swelling would go away, the fever would subside.

"Loulou, qu'est-ce que tu as?" my mother asked me as she noticed me struggling to put on a new pair of shoes. I was getting dressed to attend my friend Celia's wedding. But I barely had the strength to slip on the floor-length dress I had purchased for the occasion—my first evening gown. It hung loosely around me, though incongruously, my shoes felt too tight. My ankle, my foot, looked distended, and when I tried to walk, it hurt to take even a few steps.

Would I be able to dance tonight?

You should see a doctor, my mom said pensively as she ran her finger lightly over my ankle. She asked me how long it had been swollen. I shrugged: "A couple of weeks. Maybe a couple of months."

Mom frowned and turned to my father, who sat in his armchair in the living room, buried as always in his prayer book. Pouspous Jaune, my American orange tabby, was curled up on his lap. As my mother spoke, he didn't look up but continued to read silently from a frayed, tattered prayer book with a red cover, one of dozens he kept at his side, on a small rolling table piled high with the remnants of his lost life.

My mother glared at the cat, who gazed placidly her way. *Pauvre* Loulou is sick, she told Dad. Could it be Cat Scratch Fever?

I couldn't believe I was back to being *Pauvre* Loulou, or that we were talking about a malady that hadn't been mentioned in years. Mom, of course, was only voicing out loud what I secretly dreaded.

She described some of my symptoms to my dad as she shooed the startled cat away. Pouspous Jaune meowed as he scurried out of the room. We hadn't even consulted a doctor, yet she was already speculating that this poor creature was to blame.

My father, who listened without asking a single question, merely said that she should take me to Maimonides—our local hospital, not the Cairo shrine. Then he resumed praying. It was always time to pray in New York, though there were fewer and fewer people with whom he could pray, rarely the quorum of ten men, or minyan, needed to conduct a proper service. Once overflowing congregations were nearly all shuttered.

We were still in the cramped two-room apartment we had rented after the Cagno fiasco. I didn't have my own room anymore, but slept on a bed in a corner of the living room. Our family had been in America ten years, but with Suzette now in Los Angeles, and Isaac and César in Manhattan, my mom and dad wondered what they had to show for it.

My father, now in his seventies, was much frailer. On one or two mornings a week he attempted to venture out to pray, as was his wont. But finding a working congregation at 6:00 or 7:00 a.m. in our forlorn little corner of the world had become almost an impossibility, so that my father roamed and roamed the streets of Bensonhurst in search of any surviving temple where he could worship with the other old men who had also stayed or been left behind.

I watched as he tried to cross Twentieth Avenue, a tall, stooped figure with a limp so conspicuous every step seemed painful. He walked with his cane, which he raised menacingly in the air, waving it like a weapon at the cars that whizzed by him. I held my breath for fear he would be hurt or run over as he battled the morning traffic, forcing trucks and motorcyclists to screech to a halt to accommodate his deliberate pace. Miraculously, he never was. It was as if he wore an invisible shield that protected him against all odds. I would continue walking only after making sure he had made it safely across, following him with my eyes until he turned a corner.

On this bitter cold February night, as Mom and I headed to Celia's wedding, it was my turn to have difficulty walking.

Arm in arm with my mother, I trudged to the Cotillion Terrace, the gaudy catering hall where the ceremony was being held. Mom had been so rattled by the sight of my ankle, I didn't want to worry her any more by admitting that almost every step I took hurt.

The Cotillion Terrace was located on Eighteenth Avenue. To get there, we first had to walk down Sixty-fifth Street, which meant passing the considerably more modest catering hall known as La Perville. Small yet oddly radiant, La Perville catered to Christians and Jews, Italian-Americans from the surrounding blocks as well as Hasids from nearby Borough Park. Some nights, I would glimpse men in dark coats and fur-rimmed hats or skullcaps, while other times, I saw priests circulating among crowds of women in high heels and bouffant hair.

My mother loved to linger on the sidelines, gazing at the brides as they entered La Perville in all their white finery. How she strained to peek inside as guests gathered in the ornate lobby, sipping champagne from tall fluted glasses or helping themselves to hors d'oeuvres served by handsome waiters who seemed to glide along the carpeted interior. There was always a small string ensemble positioned close to the door, greeting guests as they entered with the melodies of old Capri or prewar Vilna.

The band played in front of a gushing indoor fountain, and Edith would stare and stare at them, a diminutive woman in a long blue woolen coat much too large for her, longing for the day when Suzette or I would be married, and we would choose La Perville, and she for once would find herself on the inside

On that cold night, I was the one who wanted to loiter, grateful for any excuse to rest. I wished that vanity hadn't compelled me to wear my new high-heeled shoes. A thin layer of snow blanketed Eighteenth Avenue, which made it even harder for me to walk. It cheered us to walk inside the Cotillion, a large, garish establishment that had once been a movie theater. It was decorated with plush red carpets, tall stairways, crystal chandeliers, and mirrors.

The wedding was in full swing. We were instructed to head toward the women's side of the grand ballroom. My heart sank at the realization it was going to be a segregated wedding, with women and men sitting apart—and dancing apart. There would be none of the romantic slow-dancing with boys I had hoped for on this night of my first evening gown.

Friends waved to me to join them in a hora. I felt winded after only a few steps and returned to my table.

The evening had hardly begun, and I was already spent.

Mom spotted me sitting alone. "Loulou, tu ne danses pas?" she asked; Why aren't you dancing? I pointed to my overflowing plate of food and pretended I was merely taking a break to sample the delicious food. I needed to get through only another hour or two of Celia's wedding.

A few days later, my mother and I ventured to Maimonides Hospital. There was no hope of a miracle inside this chaotic jumble of clinics and emergency rooms that catered to the indigent poor who couldn't afford a private doctor. There were only endless waits and, at the end of the wait, a session with a physician or resident who was often foreign, poorly educated, and barely able to speak English.

The young Indian resident who saw me seemed puzzled by my swollen ankle, though not overly concerned. He ordered a series of blood tests. When we returned for the result some days later, he merely shrugged, saying nothing seemed out of the ordinary.

I was feeling worse and worse. I was now having trouble merely getting up and going to school. Another trip to Maimonides was in order. This time, my mom and I opted for the emergency room instead of the unwieldy clinics. The wait was shorter, and I was seen not by a foreign doctor but by an American nurse in a spotless uniform and with a confident manner.

First, she ordered me to take off my sock and show her my ankle. Then, she asked me to remove my trousers so she could examine the area more closely. She summoned a colleague, another nurse, for counsel. Both seemed amazed to see it wasn't simply my ankle but my entire leg that was bloated.

I heard her gasp when she spotted it—the odd swelling above my thigh I'd neglected to mention to anyone, even Mom.

How long had it been there? she wanted to know. I shrugged, too tired to tell her my history with Cat Scratch Fever. Why hadn't I seen a doctor? I tried to explain about that too, how it was impossible to find a good doctor in New York, far harder than in Cairo, but she'd stopped listening.

Summoning my mother from the waiting area, she informed her that she was arranging for me to be seen at once by a specialist.

At once, she repeated.

The surgeon, a dapper and elegant middle-aged man named Dr. Reich, met me in an examining room upstairs. In his expensive suit and shiny silk tie, he projected the image of the *bon docteur.* He looked me over carefully, intently—the first time that a doctor had in years—all the while keeping up a light banter and wearing a steady smile on his face. He stopped smiling when he reached the area above my thigh. He asked my mom into the examining room and began speaking as if I weren't there.

"Your daughter is very sick," he said bluntly. "We need to admit her immediately and run some tests."

It was already Thursday. The prospect of a weekend in the hospital, away from my parents, seemed unbearable. I pleaded for time.

He reluctantly agreed, but only after my mother vowed we would be back on Sunday.

My father abandoned his armchair Sunday afternoon and put aside his prayer books. He shuffled up and down our small apartment as my mother helped me pack my suitcase. It was small and colorful and compact, not at all like the bulky brown bags piled up in our basement.

The suitcase was my prized possession—the first I had ever owned, since among the original twenty-six, no one saw fit to let me have my own bag. Suzette had given it to me as a gift several years back, with a twinkle in her eye, after making me promise I would use it to make grand voyages to glamorous destinations. Inside, she tucked in a small pink and white pamphlet entitled "You Are a Woman Now," with an illustration of a pretty, smiling young girl on the cover. It was a basic primer about the facts of life, but even with my mother's squeamish attitude toward sex, it contained nothing I didn't already know. As I packed, I thought of that long-ago book with its image of the young girl, looking all flushed and hopeful, "You Are a Woman Now," and wondered what had become of her.

Brooklyn didn't have taxis we could flag in the street, so we called a private car company to take us to the hospital. I rode with my father in the back of the car—he couldn't bend his leg anymore and needed the room to stretch. My mom sat in front with the driver. None of us spoke much. At Maimonides, we were directed to the children's ward, in a rotunda painted in cheery yellow and decorated with stuffed animals, toys, and flowers.

What was I doing in a children's ward? I asked. Wasn't I a woman now?

"Dearie, you don't want to be with the adults, believe me," a nurse said as she escorted me to my room and pointed out my bed by the window. After helping me unpack, she coldly told my parents it was time to leave.

My father was seated in the armchair by my bed. He had whipped out the worn little red prayer book he carried in his pocket at all times, and was deep in prayer. He wouldn't even have thought of arguing with the nurse. "Merci, mademoiselle," he said politely, and tipped his hat. He stood up painfully from the chair and shuffled a few steps behind my mom. As they waited for the elevator, he was leaning heavily on his cane.

The view from the hospital window was desolate and bare. I could see the silhouette of trees against the sky, and the faint outline of the El in the distance. I wished that my mother could have stayed by my side, as on that night we'd spent together at Maimonides' true home, the Temple of the Great Miracles, not this cold impostor that bore his name. Tonight, before leaving, she had assured me that my father would be up all night praying. Under my hospital pillow, I could feel the gift he left behind, the threadbare red prayer book from Cairo.

IT WASN'T CAT SCRATCH Fever.

After a week of subjecting me to every test imaginable, the doctors at Maimonides, like their colleagues in Egypt a decade earlier, were puzzled. They decided that a small operation was in order to examine and analyze the actual site of the swelling. They called the procedure a biopsy.

The morning of the surgery, my father performed an operation of his own. He ordered a car to take him to Ocean Parkway and the new home of the Congregation of Love and Friendship, and held a special prayer vigil to coincide with the exact time the surgery was taking place. To his relief, at least twenty men were on hand at that hour of the morning, more than enough for the requisite quorum.

Even so, the test results were dire. I had contracted another mysterious ailment known as Hodgkin's disease.

"*Hopkins* disease?" I asked, thoroughly confused. I'd never heard of it, and no one breathed the word *cancer*, though of course that is what it was.

At home, my mother seemed to unravel before my eyes. She sat at the kitchen table, penning letter after distraught letter to Suzette in California, pleading with her desperately to come home, to come home at once and help her to cope with the crisis. My father did nothing but pray, all day long and late into the night. Our apartment became his very own house of worship. I lay in bed, staring at the ceiling, staring at my leg, unable to figure out what to do.

My sister, on the other hand, was a maelstrom of activity. In L.A., she was both in close touch with us and yet mysterious about what she was doing. I only knew that she would phone at any hour of the day or night, saying I should trust no one and listen to no one. I didn't have Hodgkin's, my older sister insisted, confident as ever; all I had was a virus. The doctors were wrong, the hospitals were wrong, the tests were wrong, the biopsy results were wrong, my parents were wrong, everyone around me was wrong and not to be trusted.

She urged me to leave at once for California, where she vowed to take me to proper doctors. New York was like Cairo, she said contemptuously. Stanford in Palo Alto, the Mayo Clinic in Minnesota—*those* were the places I needed to go for care, she said, warning me not to be ministered by some hacks at a Brooklyn hospital named after a dead mystic.

I had trouble walking to the corner bakery—I wondered how I would get to Stanford. The exchanges drove my mother to distraction.

New York, May 10, 1973

Dear Suzette,
Please stop these chimeras about going to Stanford for
treatment. We don't have the money for Stanford. Loulou's
only form of insurance, if you recall, is the Medicaid card
your father helped her to obtain. If we can't decide on a
course of treatment, we will surely lose her. Maman.

Occasionally, my sister offered advice that seemed to make sense. When she advised us to get rid of the cat, Pouspous Jaune was immediately banned from the house. For days, even weeks, thereafter he would try to return, meowing at the window of our ground-floor apartment, clamoring to be allowed back in. I would see him wandering around the neighborhood and wonder how he was faring, this spoiled house cat used to being hand-fed bits of cheese by my father, and now forced to scrounge around for his supper. He seemed so lost—as lost as I felt.

My mother was resolute, and never let him in the house again. He was the last Pouspous I ever owned.

My family located a specialist, a Hodgkin's expert named Dr. Lee. He worked in Manhattan, a place so foreign and remote my friends and I called it "the City," at a hospital called Memorial. It was a major cancer center even then, but that was before it reinvented itself and wanted to be known only by the crisper, clinical-sounding name of Sloan-Kettering, when it touted the supremacy of its laboratories and research scientists over the more human mission of its doctors and clinicians.

One morning, my parents and I clambered into a private car for the drive to Manhattan's East Side. We had almost stopped taking subways: my parents weren't letting me walk anywhere. I worried that their entire savings were now being spent on taxis to and from my doctors' appointments.

I was used to doctors with ethnic names, mostly Jewish. I couldn't get my mind around Dr. Lee's identity, which only fueled my anxiety. Dr. Lee had to be Chinese, I decided. As we crossed the bridge into Manhattan, I began chatting amiably with the driver.

"Loulou, ne parles pas avec le chauffeur," my father said in a chiding tone. I wondered how even in these desperate circumstances, he could still manage to be so class-conscious that he felt it necessary to tell his daughter to stop talking with a cabdriver.

What class did we belong to, anyway?

By my father's template, we were still members of an elite, a ruling aristocracy. He was the Captain, and I was his Egyptian princess, even though all trappings of our former life were gone, and the closest we came to royalty these days was Kings Highway, the shopping strip near Ocean Parkway where Mansoura's and other Oriental grocers were located.

The driver dropped us off at the wrong address. Bedpan Alley, as this sliver of Manhattan's Upper East Side is called, is a beehive of medical institutions, research laboratories, clinics, and medical schools. We wandered, lost and hopelessly confused. After going from one building to another, we finally found our way to Memorial's lobby.

We were still early, and my parents were anxious for me to have lunch. Ever since they'd realized I was losing weight, they had become obsessed, my mother in particular, with my diet. She'd push large plates of food on me.

Memorial's cafeteria was minuscule, more like a take-out counter. Among the few offerings on the sparse menu was vegetable soup, and I agreed to my father's offer to purchase a small bowl for me. Once I began stirring the hot broth with my spoon, I noticed small chunks of meat swimming alongside the celery, carrots, peas, and onions.

I realized at once that I couldn't eat it.

"Ce n'est pas kasher," I cried. I pointed out the pieces of forbidden beef to my dad. I was sure he would be as upset as I was. Hadn't he railed against my siblings for having abandoned the Jewish dietary laws shortly after we arrived in New York? In all the years I had known him, the opulent years, the struggling years, the desperate years, the years of exile and flight, the years of personal and financial ruin, the years selling ties in the street and the years seated in an armchair at home, I'd never once known him to cut corners, to sidestep the faith that was the centerpiece of his life.

"Loulou, manges," he said very simply; Please eat.

He had reverted to that eerily mild-mannered tone he used to convey only what was most important. As I pushed the cup of soup away, he pushed it back to me and nodded his approval.

I took a sip of the broth. I felt unbearably sad; only at that moment did it finally dawn on me how sick I must be.

At last, it was time to go upstairs for my doctor's appointment. When the receptionist called out my name, I was taken to a small examining room and asked to sit on a table. I wasn't even expected to change into a dressing gown but could remain in my own clothes.

After a few minutes, the door opened and a tall man walked in, wearing dark gray trousers and a blue cotton shirt with the sleeves rolled up. He looked nothing like a doctor. He wasn't wearing a white coat. He didn't carry a stethoscope. And he wasn't Chinese. When he reached out to shake my hand, he introduced himself as Burt Lee.

Dr. Lee didn't say much, which added to my confusion. He seemed cold and somewhat forbidding. There were none of the pleasantries

Dr. Burton J. Lee III, M.D.,
of Park Avenue, Yale, and
Memorial Sloan-Kettering.

doctors engage in, not even the obligatory smile, but that was fine, since I didn't feel much like smiling either. I noticed that he was eyeing me very carefully, taking in my hair and face, even studying my clothes. I looked a sorry sight in the baggy brown pants and pale blue T-shirt that had become my daily uniform. Who needed to dress up for a doctor, anyway?

If he held back on some of the niceties, he also spared me the endless questions I had been asked in recent weeks. As doctors struggled to make a diagnosis, they would confront me with a litany of tedious inquiries. With Dr. Lee, I didn't have to say whether I had lost weight, or had trouble sleeping, if I felt weak or tired, or had noted other troubling changes in my body.

He seemed able to tell at a glance that the answer to all of these questions was yes.

Instead, as he examined me—rapidly, with a sure, firm, confident hand—he asked questions no other doctors had. He wanted to know what I liked most about school and if I had any hobbies. Was I planning to attend college? What were my favorite books and authors? He almost had me chuckling as he spoke about attending Yale in the 1950s, and going to Vassar, where I had been accepted, to meet women.

I had only one question for him: "Do I have Hodgkin's?"

"You might," he replied, as lightly and noncommittally as if I'd asked him whether I had a cold. "You might."

We returned a few days later. As I waited for him in the examining room, he met with my parents in an adjoining office. After a few minutes, I heard what sounded like a scuffle. I opened the door a crack.

There was my father, tears streaming down his face, pleading with Dr. Lee. "S'il vous plaît, Docteur," he kept saying, "s'il vous plaît, monsieur." It was the first time I had ever seen my father cry. He sounded desperate and submissive, and I had never known him to be either.

But pleading for what? I had no idea what was going on, only that the formidable man in rolled-up shirtsleeves looked angry. Burton James Lee III of Yale, Park Avenue, and Greenwich seemed troubled by the display of this old man in a shabby raincoat and straw hat.

"This will not do, sir," he said in his most patrician tone, and ushered my father and mother to the reception area.

Dr. Lee came to see me in the examining room. Without fanfare, he began to describe some of the tests he was ordering. I noticed that he used the same mild tone of voice that my dad used when he talked about grave subjects. He only sounded urgent when he leaned over to deliver one piece of advice. "Don't listen to your father," Dr. Lee told me. He repeated the warning in his clipped upper-class English, enunciating every single word: "Don't listen to your father."

And so, in the middle of my fears about my illness, and the perils the treatment would bring, and whether I would even respond, and if I would ever feel like myself again, I wondered what on earth this strange doctor—who didn't look like any doctor I'd ever known—was trying to tell me.

Why did he want me to steer clear of Dad? Why not warn me about my mother, instead, who constantly feared the worst? Or Suzette, who kept insisting I wasn't even sick?

I began going to Memorial every day for a flurry of tests. There were countless X-rays to be taken. I would be in and out of changing gowns, hustled into dark cold rooms with gargantuan metal machines I had never seen before. I took tests that lasted minutes and tests that took hours to perform, and made me ill. Everywhere I went, I learned to identify myself as a "clinic patient," and flash a plastic ID card that identified me as such.

I realized early on that in the world of Memorial Hospital, there were two kinds of patients, private and clinic, wealthy and poor. The private patients had a special plastic card and enjoyed all kinds of amenities, the most important one being access to a single physician for their care. The clinic patients, most of whom were covered by government programs such as Medicaid, were assigned a senior supervising physician, yet in my eyes, they seemed to be at the mercy of whichever young doctor or fellow picked up their chart when they showed up for their appointments.

I noted one more odd class distinction. Each day, patients were expected to have their blood drawn for a count of red and white cells. As a clinic patient, I had been directed to a second-floor laboratory that insisted on taking a full tube of blood from my arm. But private patients could have a simple prick of the finger that yielded the needed drop of blood that was then smeared on a slide; the procedure was done in the modern efficient fourth-floor laboratory. I learned to appeal to the amiable laboratory staff to have them do the pinprick and spare me the elaborate blood draw.

There was also extreme kindness from the most improbable quarters, and a sensitivity to our financial plight. In a culture that would change drastically at Memorial and other hospitals, there seemed to be little concern about payment. A billing clerk was always trying to reassure me: "Worry about getting well," he would say, "not about the bills." At the end of each treatment, another staffer would call a taxi, and give my parents the money to pay to take me home to Brooklyn.

I used most of my wiles to maneuver into being seen by Burton Lee. I'd ask in advance what days and hours he'd be on duty and make appointments only for those times. At the reception desk, I'd brazenly assert, "Oh, I am a patient of Dr. Lee," and if he was busy, I'd say that I was happy to wait. I always held my breath for fear I'd be told I had to be seen by some other doctor. I knew, intuitively, that I had found my "Bon docteur."

With my medical treatment at Memorial set, my parents still felt they had to take some more steps to absolutely guarantee my well-being.

In Cairo, faced with similar circumstances—a sick daughter, possibly in mortal danger—they had known exactly what to do. I had been scooped up and taken around to visit every holy shrine and mystical enclave in the ancient city. Every rabbi, living or dead, had been summoned to help in my recovery. Any prophet known to have resided or passed through Egypt—Moses, Maimonides, Jeremiah, Elijah—was beseeched to come exert his magical curative powers.

There were no shrines here in America, or none that we knew of, and mysticism seemed in short supply. The prophets were invariably false.

Early one morning I awoke to find my parents at my bedside. Standing near them was a stooped elderly man, a hunchback, who looked so ancient he could have been in his eighties or nineties. It was Rabbi Halfon, they said, a reverential tone in their voice, and he had come to heal me. The rabbi of the Congregation of Love and Friendship was said to possess mysterious powers. If anyone could intervene with God on my behalf, my parents whispered, he could.

The rabbi, who couldn't have been more than five feet tall, put a gnarled hand on my head and bent over me as I lay in my pajamas. He began to chant out loud and mumble a series of blessings. I had no idea what he was saying, but I noticed that my parents didn't dare sit down; they stood apprehensively in the background and didn't utter a word.

At last, the holy man was done. He handed me an amulet with Hebrew scriptures, and said I was to sleep with it every night of my illness. Then he grabbed his cane and hobbled out into the chilly spring morning.

My parents seemed much calmer after he'd left. My father eased himself back in the armchair and began to pray.

CHAPTER 22

Olives

There was an invisible barrier separating the reception lounge out front where my parents could sit comfortably and wait for me on padded armchairs, and the inner sanctum in the back where I would meet privately with Dr. Lee, and allow him to examine me, determine my treatment, chart my progress, and see how I was faring.

I was now in full survival mode. I'd instinctively realized that my new American doctor didn't much care for my old immigrant parents, especially Dad, looking over his shoulders. I didn't exactly understand why, only that—for now, at least—I needed Dr. Lee more.

This imperious and formidable WASP doctor who seemed immune to my father's tears, yet was increasingly receptive to me, had to be wooed and courted at all costs, and I agonized that the slightest misstep on my part—too great a show of emotion, too much melodrama or maudlin self-pity—would prompt him to drop me as his patient. I would still be treated at Memorial, but most likely pawned off to the legions of young oncology residents and fellows who were assigned to work on the poor.

I wanted none of them. I only wanted Dr. Lee.

That meant I had to immediately shed my Egyptian sensibility and reinvent myself as an American. I had to be as cold and confident, as sober and unsentimental, as Dr. Burton J. Lee III, or as I imagined him to be.

I began to cultivate him, gauging exactly how to talk and behave with him. I prepared intently for my appointments—no baggy pants and inelegant T-shirts anymore. I forced myself to dress up, to act poised and upbeat, though I wasn't feeling much of either. And I didn't cry, no matter how sad I felt, no matter what news he delivered.

Even early on, examinations rarely took more than a minute or two. He would peer at me intently the moment he saw me. I would notice him taking in my eyes, my hair, even my clothes and my shoes. How odd, I thought. I still hadn't gotten over the fact that he didn't wear a white coat like all the other doctors I knew, and now I had to get used to his decidedly unorthodox approach.

When we talked, it was rarely about medical matters. Mostly, we chatted about books, his favorites and mine, and hobbies, and I quickly learned of his pet peeves—and he had many, from Frank Sinatra to the emerging feminist movement.

How could he possibly know how I was doing without conducting a thorough exam? I asked him one day.

"By looking into your eyes," he replied.

The test results came back. My illness, I learned—though not from Dr. Lee—was advanced, spread far beyond the site of Cat Scratch Fever. I had secretly glanced at my chart, which said that out of four possible stages of Hodgkin's—the fourth being the worst and hardest to treat—I was in stage three.

Dr. Lee never discussed my prognosis; he never even used the word. He didn't cite odds of survival or years to live. He simply spoke of the need to begin treatment, which he minimized by saying it would consist of a couple of weeks of radiation. He made it sound as simple and innocuous as taking a course of antibiotics. He warned of only one major consequence—I would never be able to have children.

There was a way to avert that, a simple one-hour operation that could potentially work. My family seemed too dazed by the avalanche of bad news to help me decide. My mother, having raised me to avoid the drudgery of kids and housework, couldn't suddenly turn around and sing a different tune. Dad, of course, had a lingering terror of surgeries; he had never forgotten his own ill-fated operation, the pounding of the hammer. Neither seemed capable of having a rational conversation with me, of expressing concern or sorrow, let alone advising me clearly as to what to do.

Out in California, Suzette couldn't have been more distraught, though not about the latest revelation. I most certainly did not have Hodgkin's, she continued to say over and over, even when there wasn't the slightest doubt anymore. The entire medical establishment was mistaken, she said. I had to ignore them all. And I certainly shouldn't subject myself to any operations that could be dangerous and, besides, wouldn't work. Short of objecting to whatever I was being told, she never offered any alternatives, beyond urging me to get on a plane and fly to Stanford.

My brothers, oddly, were the only ones willing to weigh in decisively: I should have the operation at once.

It was the spring of 1973, *Ms.* was the most talked-about new magazine, and Gloria Steinem and Betty Friedan were holding court about what made women happy—and that wasn't the traditional route of marriage and a family. On the radio, Helen Reddy was singing the anthem to the burgeoning feminist movement, "I Am Woman, Hear Me Roar." There was a revolution under way, every bit as powerful and wrenching as the one my parents had lived through in 1952. A social movement was transforming women's lives, even as my own life was being transformed by illness.

I made the mistake of thinking the two were linked, and that I could apply the lessons of one to the other.

I didn't yet have my father's wariness toward revolutions and all they promise. I found the rhetoric comforting, an antidote to my fears, a way to escape my sorrow about the course I was choosing, or that fate had chosen for me. I ultimately passed on the surgery, and started treatment.

As I walked to my first radiation session, I was determined not to cry. I was my father's daughter: I would never let anyone see me break down.

AT HOME, MY FAMILY behaved so strangely; that was the worst of it. They were overly solicitous, which only fueled my paranoia. How sick was I, anyway? Was it perhaps worse than I knew?

Once, when we came back by subway after a long day at Memorial, they insisted on calling a car service to pick us up from the train station. We were only a couple of blocks away from the apartment. I was so distraught, I refused to get in the car and began walking home without them.

The more attentive they were, the more I fretted.

It was almost as if they didn't think the treatment was going to work. As far as Mom was concerned, I was Baby Alexandra.

Only my father continued to act as he always had, and when the treatment became so intense I could no longer eat, it was my father who thought of olives.

He began to bring home cans of olives, black, with no pits—the enormous ones labeled "Colossal." I couldn't bear to look at food, the mere thought of eating made me ill, but I was somehow able to nibble on olives. Dad insisted on feeding me himself, one olive at a time. While all around me, patients became weak from the treatments, collapsed, dropped out, never to be seen again, I continued to eat olives.

My mom, panic-stricken at my weight loss, persisted in preparing massive hot Levantine suppers night after night—stuffed artichokes, meatballs, veal stew, cooked lamb—all the dishes I had loved and now couldn't go near. She looked so hurt each time I pushed a plate of food away, unable to swallow a single spoonful of the rich stews and soups she served me and that were designed to fatten me up and make me strong.

That is when my father would step in and hand me an olive.

I became convinced that the olives—not the radiation treatments—were healing me.

I don't think Dad gave much thought to the unusual regimen he had devised for me. Olives had been a staple of life in the Levant, as essential as bread and water. While the rest of us learned to devour hot dogs and hamburgers and French fries, he preferred to dine simply on pita bread, feta cheese, and black olives. Of course, he had always lived as if he were still in his beloved Cairo. Even Pouspous Jaune had been trained like her predecessors to disdain supermarket cat food and enjoy fresh pita and black olives. It was all part of my father's plan to ignore the New World and pretend he'd never left the old one.

He sat in his chair, praying, eating his bread and cheese and olives.

From California, my sister kept calling, asking questions. Was the radiation working? she wanted to know. Was I cured yet?

There were no answers that spring and summer. *Cured* was a mythical term, I learned, used only on TV and in magazines. There was only the chance that the relentless treatments—if I could withstand them— would interrupt the disease's relentless advance.

There were times I felt so weak that all I wanted to do was sleep. But then I would feel my father's gentle tap on the shoulders.

"Loulou, Loulou, reveilles-toi," he'd say; Loulou, Loulou, wake up.

He came every couple of hours to feed me olives. I'd be so annoyed, preferring my comatose state. Only grudgingly would I take what he handed me. He seemed content if in the middle of the night, I agreed to have one or two olives. In the afternoon, around lunchtime, I was supposed to eat a few more, maybe five or six. They were regular feedings, delicate, as with a newborn. He attended to me as patiently and meticulously as my grandmother Zarifa ministered to my cousin Salomone in Cairo, in 1944, bringing him regular portions of bananas and raw eggs and stewed apricots—above all, *mesh-mesh*—to cure him of his pleurisy.

In July, deep into the treatment, Suzette surprised us by flying into New York and swooping into the house on Sixty-fifth Street. I was jolted out of my lethargy—the deadening routine of going every day to Memorial for the radiation sessions and coming home ill and exhausted. I was thinner than ever and more frail than ever. I stared at my sister, all beaming and resplendent in a kelly green sweater that set off her long, shiny black hair, as if she were a visitor from another planet.

I should have been used to her fact-finding missions, the trips she took once a year to check on how we were doing and insert herself into the latest family drama. She would come with an armful of gifts, and I tended to remember her visits mostly by what she had brought me on a particular year—games, lavish clothes, expensive chocolates. This year, her self-styled role of Inspector General had taken on a grimmer cast. She still harbored doubts about my chosen treatment, and hadn't given up hope I'd come to my senses and fly off to the Mayo Clinic. Clearly, she had been alarmed by my mother's letters and had decided to come see for herself how "*pauvre* Loulou" was doing.

And maybe she simply wanted to cheer me up.

At one point, she asked me why I hadn't bothered to comb my hair. I was afraid to tell her the truth: that if I brushed it too much, some would fall out. Until now, it had been my secret. I had very long hair, and it was only falling out in the back, where I'd been heavily radiated; I figured that if I arranged it artfully, no one would notice. Suzette took out a hairbrush from her handbag and began to lightly graze my hair with it. We both pretended not to notice the thin strands that landed on my shoulder.

In the midst of the inevitable discussions about my treatment and the mentions of Stanford and the Mayo Brothers, Suzette announced that she was taking me out to lunch. Nothing my parents or I said could stop her. I insisted that I wasn't hungry, but she was adamant. We took a subway to Grand Central and walked over to the Pan Am Building. She had made reservations at a restaurant called La Trattoria, a vast, jaunty space that was supposed to make customers feel like hopping on the next Pan Am flight to Rome or Milan (far more desirable destinations, I thought, than Palo Alto or Rochester, Minnesota). Over a lunch of eggplant parmigiana, I felt oddly chatty, though I didn't eat much.

All around us were young businessmen wearing suits, and I found myself staring at them. Elegant men in dark pinstriped suits weren't a common sight in my corner of Brooklyn. I'd point them out to my sister, remarking on how wonderful they looked. They seemed rich and important and I told my sister the man I'd someday marry would have to wear a terribly expensive suit.

At that moment at La Trattoria, as I nibbled at a breadstick, "someday" seemed like a possibility.

After lunch, we wandered around the corner to a boutique and my sister encouraged me to go in. She went to a rack and pulled out a knit sweater in a shocking shade of red—the same shade as the one I'd worn that long-ago night of the red dress. It was clingy with a plunging V neckline.

"You don't think it's too low cut?" I asked Suzette, fearful she would say yes. She didn't even blink.

"It is perfect. Why don't you wear it?"

I walked out in my new sweater, feeling resplendent and, for the first time in months, pretty.

As the summer wore on, I finally dared to ask Dr. Lee if I would be able to attend college in the fall. I was supposed to start Vassar, yet there'd been a dreamlike quality to my decision to enroll there, as if I were merely going through the motions, without any real conviction I'd be able to attend.

"You *will* go away to college," he declared, and his voice resonated with that booming self-confidence that in my mind was distinctly American, enhanced by the kind of absolute certainty that comes from growing up privileged and attending great schools.

Dr. Lee didn't betray any doubts at all, and because he seemed so sure of himself, and his voice sounded so brisk and refreshing, it reminded me of that first sip of lemonade from the glass my mother handed me at the end of the Fast of Lamentation, bracing and tart and delicious at the same time, so that I began to feel hopeful, and considered the possibility that I would be well again.

WITHIN MONTHS, IT WAS as if it had never happened.

There were no obvious traces of my illness, except that I was pale, and tired very easily. At synagogue on Saturdays, when the congregation rose, I was unable to stand and stayed in my chair, like the old people. During the High Holidays, for the first time in my life, I didn't fast. I was so thin, I concentrated on eating and trying to regain some of the weight I had lost.

One day, shortly after the treatment had ended, I developed a high fever. I returned to Memorial, but Dr. Lee was nowhere to be found; the emergency doctor on call prescribed horse pills I could barely swallow. When Dr. Lee returned, I showed him the pills and he ordered me to stop taking them. The fever disappeared as mysteriously as it had come.

In September, days after my last treatment, I went away to college. I left my father behind in his chair with his cat and his prayer books and his Levantine eating habits and rarely came back—a few days here and there, weekends, major holidays. Only every once in a while, I'd feel sad, and I couldn't understand why. I went home when the terrible sadness came over me, the feeling I'd first experienced Sunday night in the hospital, a sense that I was trapped and had to run, run, but where, *where?*

Off I'd go home to Brooklyn. Then, within days, or hours, I'd return to Vassar. I kept to myself. I no longer had many friends, and no boyfriends and no red dresses. Illness had made me wary: I had decided to take life's gifts sparingly, like olives, one at a time.

My family rarely mentioned my malady. As I got better, my father seemed to get worse. He rarely left his chair. I had a difficult time accepting his decline, though illness was what had always bound us together.

Dr. Lee's edict against listening to my father had long since expired, yet I was still in its grip. It was only Dr. Lee that I listened to now. I lived for my checkups with him—once a month at first, then every couple of months. It was the only time I felt safe, as if I were out of danger only there, inside that little examining room, with him at my side. He did nothing much except talk. There were no medicines to prescribe, no injections, and few tests. As long as I looked well—and he insisted he knew the moment I walked in—that was the proof the disease was under control, and we could move on to other more important subjects.

We chatted for hours, and it was as if his other patients, duties, responsibilities, were put on hold while we huddled in the small room that hadn't changed from the day I'd first climbed on the gurney, a frightened teenager in baggy jeans and a frayed light blue shirt and no hope at all.

One day, a couple of years after the treatment, I asked whether I was cured.

I had read about five-year marks and ten-year marks, yet I noticed that Dr. Lee never spoke in those terms, or even of remission. I was either well or I wasn't.

He frowned at the word *cured*. Hodgkin's, he declared in that imperious tone he'd occasionally adopt, wasn't curable.

"It will return," he warned. "It will come back."

I went home so panic-stricken I couldn't sleep or eat. And of course I wouldn't confide my fear to Mom or Dad: it was their nightmare, too. It was the nightmare that had usurped my teenage years and made it impossible for me to enjoy that most essential element of being a young girl—the sense of feeling carefree.

The next day, without bothering to make an appointment, I returned to Memorial and demanded to see Dr. Lee.

I wanted to know when "it" would be coming back. He looked at me, somewhat startled. He was an impetuous man, not always careful with his words. I could tell he was in a more pensive mood, and I had learned over the years that there was no kinder or more caring human being on earth than this tall, autocratic WASP.

He gently took my arm. "Some of us have this much to live," Dr. Lee told me, indicating the space from my fingertips to my wrist, "and some of us have this much to live," showing me the longer expanse from my wrist to my elbow, "and you don't know, and I don't know, so forget about it."

And that was the closest Dr. Lee and I ever came to discussing my "prognosis."

I never even asked him how people with my stage of Hodgkin's fared. I don't think he would have told me. Besides, from what I'd observed at Memorial, luck seemed to be the key determining factor in which patients got better.

Doctors over the years would insist that I was cured, but I would shake my head, because I knew better.

Since the summer of my sixteenth year, my life had hung in the balance, and I knew it always would. I watched enviously as my peers, young women my age, planned their futures—weddings, careers, families, vacations. I could never plan so much as a dinner a day or two in advance without the nagging sense of fear that it would never come to be. The voice that I'd first heard at the time of my illness would whisper: There will be some terrible calamity, and it won't happen, and all will be lost.

One day, when I came home, my father surprised me by starting a conversation with me, which he seldom did. He seemed oddly agitated.

He put down his careworn red prayer book to tell me about a dream he'd had about us. In the dream, he was giving me two almonds, two perfect white Jordan almonds, the kind that they'd give out on joyous occasions once in Cairo—at engagement parties, births, weddings. He used the French word for the delicious sugar-coated almonds: *dragées.*

Neither of us spoke about what the dream of the almonds meant, and I didn't think much about it at the time. But I have never stopped thinking about it, and of what my silent, noncommunicative father was trying to say to me.

What took place on that long-ago afternoon between my father and Dr. Lee? I would finally inquire years later; by then, Leon was dead, and Dr. Lee had become a friend and mentor, someone I cherished deeply and communicated with regularly. Why did my father break down that day, I wanted to know—what was said inside that office?

Dr. Lee had treated hundreds of patients since the encounter, had spoken with thousands of anxious parents or spouses, either to reassure them or to deliver a grim verdict about their loved one, yet he seemed to vividly remember his lone exchange with my father that day in the spring of 1973.

My parents had come in together, he said, and both had sat by his desk. What was striking was how silent my mother had been, how she hadn't said a word during the entire meeting, Dr. Lee recalled. She'd simply sat there nodding, a frightened look in her large brown eyes, while my father took the lead. Though always a man of few words, who had retreated into a shell of silence in the last few years, Dr. Lee remembered him talking animatedly and without pause.

His daughter was in grave danger. She needed care, the best care. If not, she would be lost. He wanted to know if Dr. Lee could take over my case.

That is what he kept asking over and over again, would Dr. Lee become my personal physician, would he take over my case?

Dr. Lee was used to fielding appeals from patients and families from all walks of life. Some years later, men close to the shah of Iran would bring him to New York and ask Dr. Lee to examine him and weigh in on his care. Later still, he would be appointed as the physician to President George H. W. Bush. He came from the noblesse oblige school of medicine—that the well-to-do have an obligation to take care of the less privileged—and that was the part of medicine that he liked most.

What he didn't like—what he minded, perhaps inordinately—was the imploring tone of this elderly man. My dad was like no one he had ever met, not simply from a foreign culture but almost from another planet. He was obsequious to a degree that made him cringe. Dr. Lee, who had made it a point in his career to treat rich and poor alike, who liked to think he cared for patients without regard to their wealth or social status, was taken aback and perhaps offended.

There was also the desperation in my father's voice. "This man has no cards left to play," Dr. Lee remembered thinking.

He couldn't have known how close to the truth he had come. My dad was reduced to begging because there was nothing else he could do. Faced with the prospect that his youngest child was dying, that her only hope for survival lay with a patrician American doctor he couldn't afford—with whom he couldn't even communicate—the boulevardier of Cairo had nothing left to trade: no money, no position, no social status, no white sharkskin suit.

Looking back, Dr. Lee would have to concede that he had been abrupt and possibly harsh. By ushering my dad out of his office that day, he had seemingly dashed his last great hope—that this distinguished American doctor would rescue the child of his old age, the teenager that he insisted on calling incongruously by her childish nickname, Loulou. Of course, Dr. Lee had proceeded to do precisely what Dad had asked—he had taken over my case, he had become my private physician, he had saved me—leaving me to wonder years later whether by breaking down and pleading his case like a mendicant and invoking me again and again, my father had in fact found one last card he could play.

CHAPTER 23

The Guardian of
the Orphans of Jerusalem

I t was his last apartment, though I didn't know it at the time, of course.

After the tumult of Sixty-sixth Street and the bitter taste of *l'affaire Cagno,* the enveloping sorrow of the House of Prayer left us with no choice but to move again, and again only a few doors down, because by now we felt defeated and exhausted, convinced no world beyond Sixty-fifth Street would have us, and even our existence there, as my illness had shown, was tenuous.

For once, it was the proper size—not too big, not too small, fine for the three of us.

My mother, for one, was relieved simply to be out of what she called our "bad-luck" apartment.

"Pauvre Loulou—cette maison lui a porté malheur," she kept saying; Poor Loulou, this house was unlucky.

It was as if some element of those shabby little rooms had been responsible for my getting sick every bit as much as Pouspous Jaune. No one dared to question her logic. Desperate to explain the unexplainable—why I had contracted cancer at sixteen—we insisted on pinning the blame for my illness first on a cat and then on a cramped apartment whose windows all faced a dusty courtyard in the back.

César surprised us by moving back home, and we were almost a family again. He had tired of the single life, and missed the comforts of home and the room he had shared with my dad all those years. My father was delighted to welcome him back, so there they were, roommates all over again, as they'd been when we first came to America.

I had my own room once more, small, at the front of the house, though my father didn't speak of the pleasures of watching the street life anymore. He didn't speak at all. What he did was position his green-and-white beach chair—which hadn't seen the beach since my illness—close to the window facing Sixty-fifth Street.

The chair was lined with pillows to soothe his aching back and waist and hips, and he put his prayer books one on top of the other on the small tray table my mom had purchased specially for him from Woolworth's. In the corner, within view at all times, was the suitcase he had purchased for the day he was going back to Cairo.

That became his entire world—the beach chair, the prayer books, the tray table, the window, and the small vinyl suitcase.

And the radio.

Home all day long, he was glued to the radio as he had been as a young man in Cairo, when he'd sat for hours listening to Om Kalsoum's laments.

The soothing sounds filled the living room, and instead of the Cairo Diva, I heard the mellifluous voice of "Your host, Charles Duvall, broadcasting from the shores of Lake Success."

"Where is Lake Success?" I found myself wondering. It seemed so alluring, as charmed and seductive as Duvall's radio persona. Somewhere in this world, I thought, sits a handsome man with a debonair French accent inside a studio overlooking a magnificent lake, and he is so filled with confidence and serenity simply gazing at that body of water that it seeps into his voice and his words. He proceeds to calm us all, infuse us all with his confidence and serenity.

Lake Success. It was where I wanted to be; here in America nothing else mattered.

My father would sit for hours hunched over his prayer books, usually the little red book I had given him back when I came out of the hospital. As the 1970s—a horrible, wretched decade, as far as I was concerned—came to a close, it became impossibly torn. I didn't think the book would survive another day without disintegrating in his trembling hands, the pages falling out or crumbling into dust. He had long stopped trying to repair it, so that even the Scotch tape and duct tape and masking tape and surgical tape that had held it miraculously together all these many years were all dried out, and the red jacket had turned into a somber maroon brown. The book, I realized, had *become* my father. The two even looked alike, all bandaged up, small pieces breaking all the time, both trying to hold on, both in danger of disappearing.

Sometimes I'd walk in and find the prayer book mercifully closed. What was propped open was my father's sky-blue checkbook, and he was patiently, meticulously signing checks. Around him were pieces of the morning's mail. Other than his stock statements, which continued to stream in—shareholders' notices from Zambian copper mines, or the Consolidated Gold Fields of South Africa, or Sperry-Rand, the quixotic investments of his years in America that had failed to make him rich—almost all the mail was from far-flung charities.

The orphanages and schools to which he donated money, daily and compulsively, kept in close touch. Brown packages with odd-sounding names arrived from Israel; inside, there'd be handsome brochures featuring images of large cinder-block or stone buildings, alongside photographs of young children looking anxious and troubled and filled with yearning.

The Great Orphan Home of Jerusalem for Boys, the Dispenser of Kindness Orphanage for Girls, the Institute to Uplift the Souls of the Holy, the Light of Life Girl's Academy, Girl's Town of Jerusalem Academic and Vocational School, the Trade Institute of the Voice of Jacob Our Patriarch, the Maker of Great Miracles Charity Box. There were dozens and dozens of charities, as if my father were hedging his bets, contributing bits of his meager savings to each of them, on virtually a daily basis. There were occupational schools affiliated with the orphanages that sent pictures of their vulnerable charges bent over sewing machines or learning how to make tools, there were orphan medical clinics and orphan dental clinics and orphan residence halls.

An entire universe dedicated to the care of motherless and fatherless children looked to my dad for their salvation.

My favorite was the Orphan Bride's Aid Fund. I imagined a young girl, weary of years of institutional life and with no one but other orphans for company, using my dad's slender savings from brokering bolts of white lace to purchase a white lace gown of her own, or a veil.

Pay to the order of "The Institute to Uplift the Souls of the Holy," $5; pay to the order of "The Orphans of Jerusalem," my father would write in his tremulous hand, $10. Pay to the order of "The Light of Life Girls' College," $15. Pay to the order of "The Maker of Great Miracles Charity Box," $20.

I didn't immediately grasp the purpose behind the flow of donations, whose receipts and expressions of gratitude cluttered up our mailbox.

They were for my benefit. My father had asked orphanages and charities to pray for my recovery. The checks kept flowing—to this girls' institute, this boys' vocational school—all with the explicit request that recipients effect my cure with their prayers.

They seemed delighted to comply. We were deluged with offers of bountiful blessings—special prayers by orphans who enjoyed God's ear.

"Loulou, Dieu est grand," my father exclaimed when he received a note confirming prayers had been recited.

The Maker of Great Miracles, which seemed to bear a mysterious relation to the shrine of my Cairo childhood, offered my father an amulet. Behind a large, square blue-trimmed receipt that looked a bit like a stock certificate or a high school diploma was a special prayer with instructions that it be read out loud three times: "I give this donation for my poor brethren, I give this donation for my poor brethren, I give this donation for my poor brethren, God of the Maker of Great Miracles," went the amulet. "Answer me, answer me, answer me."

My father, having seen me through my treatment, was now watching over my recovery in the only way he knew how: by pursuing a miraculous cure.

Over the years, the orphanages and hospitals, old-age homes and youth towns and vocational schools and rabbinical schools were very diligent about keeping in touch, and that was the mail that he shuffled to the small metal box in the hallway each morning to collect. He was completely homebound as the decade came to a close; that was the extent of his travels—the five yards or so from the door to the mailbox in our building's lobby.

Our house became overwhelmed with tokens of gratitude—calendars, greeting cards, certificates of appreciation, more amulets. They came in a cavalcade of colors—orange, blue, purple, sea green. I began to imagine Israel as a country of orphans, all of whom depended on my father to get by. I would go to sleep at night and dream of the wide-eyed little girls in the brochures, appealing to him to rescue them.

As if the Captain were capable of rescuing anyone.

Eager to curry favor was the Great Orphan Home of Jerusalem for Boys, which acknowledged every gift with a handsome hand-engraved certificate. "May the father of all orphans reward you with all kinds of prosperity," it stated. On the back was a black-and-white photograph of the Orphan Home's bearded founder, the saintly Rabbi M. J. L. Diskin, smiling dolefully into the camera.

Below his picture, the long-dead rabbi promised to "intercede in heaven for all who support this Orphanage."

There were different rates for these celestial interventions. A one-time contribution of $50 meant an orphan would recite the kaddish, the prayer for the dead, one time—immediately after the donor passed away. For $100, the orphan would say the memorial prayer repeatedly, every year. A thousand dollars would enable the donor's name to be engraved over the bed of an orphan. My dad chose the more modest $5 and $10 route to God, and that was fine, because Rabbi M. J. L. Diskin still smiled sadly from his heavenly photo studio, and promised to do what he could on our behalf.

The Guardian of Life Orphanage for Girls was perhaps the most appreciative. It sent along a small pistachio-green book, complete with a calendar and a list of all the benedictions the children would be immediately conferring upon us. "You will be rewarded with bountiful blessings for good health," the green book vowed.

As I flipped through the calendar, I noticed that Dad had made small notations next to certain days and months of the year. They were the dates marking the passing of his mother and his father and six of his nine siblings, all carefully circled. There was my aunt Leila, in July. My grandfather Ezra was remembered a week later, with only a one-word notation, "Papa." I found it strangely jarring that as he turned eighty, my dad still called his own father "Papa," like a little boy. My grandmother Zarifa, "Mama," appeared one week after that, next to a note about his sister, Tante Rebekah. My tragic aunt Ensol, killed along with her husband, had an entry in November, as did Oncle Joseph, the oldest of the ten children. In one cruel month straddling January and February, my dad noted the passing of his two favorite siblings, Oncle Raphael and Oncle Shalom of the clubfoot and the humble demeanor and the gentle heart.

Two siblings were absent from Dad's ledger of memory: Bahia, who had perished at Auschwitz and whose date of death had never been learned, and Salomon, the priest and convert who had indicated on his résumé, on file with the monastery at Ratisbon, that he'd wanted my father, along with Oncle Raphael, to be notified in the event of his passing.

There had once been ten, and now only he and his little sister Marie were left, and he hadn't seen her since 1956. Yet he continued to remember and pray for all of them and to memorialize them in the little green book of the dead. He seemed content simply making out the small checks. It became a full-time job. The checks were for the same amounts that he had written month after month over sixteen years to pay back the debt for the *Queen Mary*—mostly $10 increments, occasionally a little more, occasionally less. The sums were deceptively small; he wrote so many checks, day after day, that he was actually giving away a significant share of his impossibly small income.

César, who worked as an accountant, worried like a wife whose husband gambles with the grocery money. My father reassured him, but kept on as before. It was his calling, now, every bit as important as selling ties had once been, or brokering the sale of yards of brocade or trading stocks at *la bourse*. In a culture of ambition and greed, my father was, as always, resolutely against the grain. He had become the

Dispenser of Kindness, the self-appointed guardian of the orphans of Jerusalem.

Officially diagnosed with Parkinson's, his hands trembled more than ever, so that the amounts he made out and the names of the objects of his largesse were almost illegible.

I felt so much better, I didn't even stop to consider how he was faring. Nor did I give him and his otherworldly approach much credit for my miraculous recovery, the fact that in the course of my continued checkups with Dr. Lee, my physician marveled at how well I seemed.

In my father's case, Charles Duvall's dreamy mantra, "From the shores of Lake Success," sounded increasingly distant and remote, as if Dad were a passenger on a boat floating farther and farther away from those desired shores.

He was not well. He was descending into a physical and mental purgatory. But he was so used to keeping silent, to being stoic about his travails, that now that he needed us, needed us to mount an intervention to rescue him in the way he had summoned the orphans of Jerusalem to save me, he didn't know how to request it—demand it—of us, his children.

One morning, he called me at work. It was unusual—he never phoned at the office, and it was as if, years later, he still hadn't made his peace with my decision to find a job and support myself instead of heeding his counsel to find a man, a rich and powerful man—*un banquier*—to look after me. Who ever heard of a woman working?

"Loulou, je ne me sens pas bien," he said; I don't feel well. He spoke so softly I could barely hear him. I listened, a tad impatiently. I had so much work to do.

"Loulou," he repeated, "je me sens très mal." I feel very bad.

I'd try to look in on him later, I promised, and hung up. It was the dawn of the Me Decade, and by focusing obsessively on work and my own needs, I was acting out its distorted values, values that had nothing to do with the far more compassionate underpinnings of my Cairo girlhood.

Like my siblings, I too had drifted. Even holidays like Passover, once so sacred, a time of waiting for Elijah, were now an afterthought. I observed them only in the most careless and minimal fashion. There were no more candlelit expeditions through the house in search of crumbs, and no sifting of the rice. I barely cleaned my apartment and usually celebrated the Seder meal itself in someone else's home, not my own, or in a restaurant.

Except that once in a while, I'd find myself yearning for those little Cairo spoons, and the musical sound they made as Dad tapped them against his wineglass. I had lost track of the steel box where they were stored, had long stopped wondering what had become of it, and the little spoons, and all the other treasures within it.

Only by chance did I learn of its fate, when it finally occurred to me to ask what had become of the box that housed so many of my childhood illusions.

A mysterious fire had raged one night through the basement and claimed the twenty-six suitcases, and all that had been so carefully arranged inside them—the handmade clothes, the brocade, the women's lingerie, two dozen pairs of a child's flannel pajamas, and saddest of all, the dark silvery box belonging to my two grandmothers, Alexandra of Alexandria and Zarifa of Aleppo. The delicate teacups and saucers, the glasses wrapped in tissue paper, the silverware, the spoons—all of the fragments and mementos of our former life were gone.

The blaze had occurred when I was living away from home, and my siblings had long since left, and no one was around to help my parents cope. Leon and Edith had never mentioned their loss. What did it matter, anyway? they must have thought in their loneliness and despair. A lot of old fineries that meant nothing to anybody anymore, and certainly nothing to their distant, assimilated, self-absorbed, and thoroughly Americanized children.

CHAPTER 24

Psalms for My Father

"*Loulou.*"

I could hear my father the moment I stepped off the elevator. He was all the way down the hall, yet he'd already spotted me.

By the late 1980s, Dad was said to have dementia and Alzheimer's and Parkinson's, yet I always doubted the diagnoses and the doctors who rendered them, especially at moments like these, when I saw how alert and clear-minded he was—his green eyes shining, his mind as vivid and intense as ever. Nor did he have any trouble recognizing me—on the contrary, he seemed to live for the times he saw me coming.

He had been stripped of any identity by then—no longer the man in the white sharkskin suit, the boulevardier, the Captain, or even the exile. He was only a patient, one of several hundreds, at the Jewish Home and Hospital, a place that was neither a home nor a hospital nor especially Jewish.

Situated on New York's Upper West Side, it was an institution similar to thousands of others: vast, cold, modern, and, to my father in his final days, unspeakably cruel. Bewildered, confused, desperate, he still nourished the hope that someone would come to rescue him. That is why whenever he saw me, he would begin to cry, "Loulou, Loulou."

Hearing him shout my name, I would start running, running down the long corridor past the other old men and women in wheelchairs until I saw him.

I would find him by the last room, a thin, lonely figure in a light cotton gown, his red prayer book in his hands. He found comfort and safety in the red book and the chants and incantations it contained, and he would mutter them to himself again and again.

I tried to embrace him, reaching for his thin, skeletal frame barely covered by the blue nightgown, but more often than not, he was too agitated. "Loulou, où je suis?" he'd ask; Loulou, where am I? And then, as some nurse passed by, he'd try to catch her eye and say with that tony British accent he still maintained after all these years, "I want to go home, please take me home." More often than not, the nurse would simply keep walking.

He had survived exile from three different countries, but it would take the fourth, America, and its quintessentially American institutions to defeat him.

The Jewish Home sparkled with modernity. It was possible to be tricked at first by its sleek appearance, to be taken in by its elegant lobby and well-heeled staff, to trust its glowing reputation and revel in the spacious visitors' rooms and gift shop and large, showy fish tank.

How I came to despise that fish tank. When I saw my father become painfully emaciated, develop ulcerous sores and countless other infections, afflictions, and maladies, I wondered why on earth an institution would lavish better care on its fish than on its patients.

I complained, of course, but to no avail. I, too, had lost my identity: I was now simply "the daughter," which meant that my objections or appeals didn't have to be taken seriously, that they could be safely ignored. And what could I say about the food, which didn't even adhere to Jewish dietary laws? For the first time in his life, my father was being forced to eat food that wasn't kosher—an abnegation of all he believed.

There was no one to whom either of us could turn.

Edith was also grievously sick by then, the victim of multiple strokes that rendered her mute and immobile and, if possible, even more help-less than my father. She, too, lived in the nursing home. Felled by a massive brain hemorrhage one spring day in 1988, she had never fully recovered. The woman with the luminous mind who had captured the heart—and the key—of Madame Cattaoui Pasha was now confined to a wheelchair, her memory and wondrous intellect all but erased.

As for their children, we were at war, incapable of agreeing about their care, incapable even of communicating.

The battle lines had been fiercely, brutally drawn. On one side was Isaac, the most Americanized of us all, who hired lawyers and doctors to have my father declared incompetent, appointed himself his legal guardian, and institutionalized him. In the process, he had displaced César, Dad's natural guardian, the son who had shared a room with him all those many years.

On the other were César and I, trying to pick our way across the nightmarish landscape of hospitals and nursing homes and offer our father some relief. Suzette was living in London—her most recent stop in the restless journey that had begun with the flight from Sixty-sixth Street. She was both removed from the fray and oddly involved, mak-ing her views known from across the ocean, the way she had inserted herself from far away in my illness.

Occasionally my father's problems seemed too urgent even for the nursing home to ignore, and off he'd go by ambulance to Mount Sinai, a large, equally impersonal medical center situated across the park, on Fifth Avenue. He was desperately ill by then, but the repeated trips to the hospital did little to make him well. Lost in a sea of beds, he barely survived treatments that were in some ways worse than his maladies, mostly because health care in America in the late 1980s and early 1990s had become so dehumanized. I would arrive in the morning and find him in one hospital room, and return in the afternoon to find him in another. By the following morning, he'd been moved to another, and to another after that.

"Where is my father?" I'd ask the clerks at the Mount Sinai nursing station. They'd check their records and coolly inform me that he'd been "transferred."

Why? I'd ask. There was never a clear answer.

I never saw him without his tattered red prayer book in hand. He would be praying even when the deck was stacked against him, he prayed even when doctors had either given up or didn't care, and his family wasn't able to do much, and there was no hope left at all.

He was praying for a miracle, of course; he never ceased believing in the possibility of one.

One day, both my parents were admitted to Mount Sinai. They arrived separately—hadn't they always?—and ended up in different rooms, in different wings, on different floors. I arranged a reunion at a patient lounge, located in one of those airy atriums that look so appealing to the outside eye. There was my father in his large rolling faux-leather E-Z Boy chair and my mother in hers.

The two looked at each other and then looked away. They said not a word, unable even to acknowledge the other's presence. It would have meant acknowledging the horror of their condition, their absolute inability to help each other. I have never felt so sad, and I suspect, neither did they.

I am sure that in these times, my father wished he were anywhere but here, in these scrubbed, soulless rooms where few people ever came to check on him simply to see if he was comfortable or in pain, hungry or thirsty, to give him some modicum of human solace. I am certain that Dad would gladly have traded Mount Sinai and the Jewish Home even for shabby little Demerdash, the public hospital for Cairo's poor—anywhere but these gleaming New York palaces of pain.

At the Demerdash, at least kindly, engaged Dr. Khatab, his surgeon, had come by every day to check on him and offer reassurance and support.

When my father was admitted to Sinai late in 1992 for an operation that should never have been permitted given his advanced age and frail state, there was no Dr. Khatab to offer comfort. I called the surgeon to beg him not to perform the operation; he didn't take my call. Afterward, if he stopped by, I never saw him. Instead, each day I would encounter a procession of earnest, pale-faced surgical residents who seemed completely removed from the patients they were treating. My father was simply a "case," one of many on their lengthy rounds.

Not surprisingly, when Dad took a turn for the worse, no one noticed

Edith and Leon at the 1964 World's Fair in Flushing Meadows Park.

until it was too late. He was moved to intensive care, and there he stayed, attached to a respirator. His entire body was failing, and yet, to the end, he still fought, hanging on for days and weeks.

I was back at my office one Friday afternoon in January when the call came. The Captain had died. Seated at my small cubbyhole, in the newsroom of the *Village Voice,* I began to scream. I thought that I would never stop screaming. "You can have a little job," he had told me once. "You can open up a flower shop."

I caught a taxi to Mount Sinai and arrived to find my father's small cubbyhole in the ICU being cleared of its tubes and bedding and any trace of him.

The funeral was held two days later. It was bitter cold, and I'd forgotten to wear a coat. Suzette, stranded in London, didn't attend. My mother didn't either; César and I didn't have the heart to tell her that Leon had died.

I am not sure if she would have understood. She had come to live with me, but she couldn't speak or swallow or move her arms or her legs. She was, at the end, only able to mouth a single word. Asked if she felt fine in the upstairs of my duplex where I'd set up her hospital bed and IV poles and small portable respirator, because she couldn't even breathe on her own, my eloquent and literary mother managed only to reply: "Okay."

EXACTLY ONE YEAR LATER, I returned to Brooklyn for Leon's memorial. It was being held at the Congregation of Love and Friendship— the brand-new one on Ocean Parkway—where each Saturday, a group of old men of Dad's generation, and a handful of young ones, gathered to read out loud the psalms of King David.

The building stood at a corner of the boulevard that had filled my mother with such longing. "Ah, une maison sur Ocean Parkway," she'd say dreamily; a house on Ocean Parkway. The area was thriving— crammed with million-dollar homes and immigrants who had become blue-jean kings and discount-chain magnates and electronic-store czars.

It was customary here, as it had been in Cairo and Aleppo, to honor the person who has died by chanting all 150 psalms in one marathon session. With each and every psalm that is recited out loud, the soul of a loved one is said to rise higher and higher, until it finally reaches its place in heaven next to God.

On this cold January afternoon, a small group of mostly elderly men showed up in the synagogue's basement to participate in the psalm reading. A couple had known my father in Egypt and continued to pray with him in America, where he could always be counted on to complete a minyan, the requisite quorum of ten men needed for a proper service under Judaic law. Elie Mosseri, who had worshipped with my father on Sixty-sixth Street in those early days in America, approached me. He had known me as a child, he said, and he had known my father. "He would come and stay eight, nine hours a day in the synagogue," Elie told me sorrowfully.

The reading had to begin, but there was no minyan, only half a dozen men who sat clustered together around a long table. I sat down next to them. They looked at each other with alarm: under Orthodox tradition, women are never allowed to sit or pray alongside men.

Of course, this wasn't a typical service. After a brief conference, they nodded and indicated I could stay. Seated with the men, I felt like a little girl again, accompanying my father to pray, permitted to sit with him in the men's section. To me, it had been the ultimate privilege.

A young woman suddenly materialized carrying platters piled high with fruit—strawberries, kiwi, melon, oranges. She set the table without a word, studiously ignoring me. She returned with more plates containing pistachios, walnuts, cashews, roasted hazelnuts. How my father would have loved the hazelnuts, I thought, reaching for some. A man seated at the head of the table began by chanting the first psalm and then another. His neighbor, an octogenarian with merry eyes, read the next psalm and one more after that. The men went around the table, each one reading a couple of the psalms.

At first, I simply listened, not daring to join in. Finally, I asked: May I read a psalm for my father?

The men turned to one another; they had already broken one rule by letting me sit with them. Would they now break another by having me chant out loud?

"Let her say it," someone shouted. "Go ahead," said another. I began to read, nervously, haltingly. It was a long psalm, and to my horror, I kept stumbling over words. But my prayer companions were kind. They simply called out my mistakes as my dad would have done, and when I wanted to give up and let one of them take over the reading, they urged me to continue.

As the afternoon waned, the pace intensified. We had to be done with the Book of Psalms before the Sabbath ended. Each time my turn came, I took a deep breath. Yet I found that I was getting better with each passage I read; I discovered an old ease and fluency with Hebrew. We neared the home stretch, yet we'd never achieved the requisite ten men. "Maybe we will count you as part of the minyan," one of the men said smilingly.

This mournful day, when I had gone from a memorial service for my mother, dead one month, to one for my father, dead one year, took on an oddly elating cast. I had received an unexpected gift. I began to think of my dad not as old and sick, but as young and vital, walking in his white sharkskin suit. I prayed for his soul to rise, and for my mother's, too, but it was I who felt elevated.

On my way out of the service, I passed clusters of people leaving the synagogue. They were holding branches of sweet-smelling green leaves, which they inhaled deeply. They chatted amiably in Arabic and French, the charming, easygoing ways of the Levant on display, the city momentarily become a congregation of love and friendship.

Cairo, Finally, and Again—

SPRING 2005

Seated aboard the Alitalia flight from Milan to Cairo, I felt suddenly as if my father were there next to me—as if he, too, were going back on this spring day in 2005, finally and again. I turned to face my husband, who had been in the seat beside mine, but I didn't see him anymore.

I was conscious only of a tall man with long legs, one of which he couldn't bend comfortably. He was old and deeply frail, but his green eyes were shining and he was alert and every bit as excited as I was about the voyage. In my mind, my dad and I were returning together, as we had hoped.

The cry that had pierced the years had at last been heard.

Ragaouna Masr: Take us back to Cairo, please take us back.

Neatly stashed under my seat was my lone, small suitcase. It wasn't too dissimilar from the bag Dad had kept for so many years packed and ready to go in a corner of the living room on Sixty-fifth Street in Brooklyn—no bigger than a breadbox, as if he'd planned to take very little with him, and certainly not the mountains of clothes and supplies we'd brought along four decades earlier.

The chatty Alitalia pilot kept interrupting the in-flight programming to point out the sights. Genoa, Naples, the Greek Islands, the port of Piraeus, and finally Alexandria harbor, only a few thousand feet below. It was if I were reversing the journey my father and I had undertaken so many years before, the voyage that we would always regret.

We had signed papers declaring we were never coming back. The Egyptian government, hungry for Western currency and Western tourism and Western goodwill, had seemed anxious to reassure me that I was welcome to return and to stay as long as I wanted—even move back, if I wished. They spoke charmingly and with apparent sincerity, as if to suggest that our family's flight in the spring of 1963 had been due to some absurd and terrible misunderstanding they were now eager to clear up, if only I'd let them.

Fine, I thought, suddenly feeling crisp and efficient and utterly American. Let them roll out the welcome mat. I was silent throughout the entire plane ride, unwilling to share with my husband all I was feeling, more concerned with sustaining this sense that my father was next to me. I wanted to feel him at my side during this trip, to believe that he had left heaven to accompany me. And yet for him, heaven had always been here—wandering through the streets from morning to night, being greeted by friends and even by strangers in a city that enveloped you, devoured you, consumed you with its love.

Stepping off the plane, I saw several people holding up white placards with my name. Representatives of the Egyptian government had come to welcome me back to Cairo. They all seemed puzzled, and perhaps also troubled, by my first request: Malaka Nazli—I wanted to drive immediately to Malaka Nazli.

It was the first place he would have wanted to visit, and the last place he had wanted to leave.

Why call the modern thoroughfare by a name that hadn't been used in decades, they asked, frowning, the name of a long-dead queen? And why, of all the wondrous sights to see in Egypt—the Pyramids, the Sphinx—was I insisting on visiting a street famous only for its smog and maddening congestion?

I insisted that nothing else mattered to me but Malaka Nazli. I repeated the name for emphasis: Malaka Nazli. I said it at every opportunity, like a child in love with a singsong.

Ramses Street, as it was now called, was impossible at this time of day—or any time of day: we'd be stuck in traffic for hours, my driver wailed. Better go to the hotel and rest for a while, he counseled amiably. But then when could I see my street? I persisted.

He huddled with his fellow drivers. Try after midnight, they said, laughing.

What he was willing to do was to drive me downtown. He offered to take me to the Gates of Heaven, the temple where my parents had been married more than sixty years earlier. And Groppi's, with the pebbled garden where all my childhood hopes had grown. Anywhere but to Malaka Nazli.

I shrugged. "To the Gates of Heaven."

As we turned the corner on Adly Street, I spotted it immediately—the immense, hulking structure with its faded stonework, its wrought-iron gate, and the delicate carvings of palm trees along the front, symbol of the Jews of Egypt. In front of the synagogue stood a small army of security guards, brandishing an assortment of weapons, including guns and rifles they pointed menacingly our way. I wanted us to slow down so I could get a closer look, but that only prompted the officers to swarm around us and warn us to keep moving.

My driver obligingly continued his tour of downtown. The streets where my father had once bought me exquisite outfits now featured cheap, hopelessly tacky merchandise. There was one discount store after another, as if Cairo had turned into an outsize version of Eighteenth Avenue, filled with merchants hawking bargain fare to customers who could barely afford them.

On top of the storefronts I glimpsed once-grand apartment buildings that looked as if at any moment they could come tumbling down. Even here, in the heart of the business district, there were clotheslines, with shirts and socks and bedspreads and lingerie flapping in the breeze.

Where was the elegance my parents had pined after? The fine boutiques and lavish, abundant department stores that carried such distinctive merchandise that we would later find ourselves disdaining the offerings of Paris or New York as inferior to what we had once known?

No one could compete with the merchants of downtown Cairo and their dreamy wares—Benzion, where we bought yards of the softest white cotton, cut and ready to be hemmed into sheets and pillowcases; or Hannaux, so snobbish, featuring the most expensive bags and accessories. And Cicurel, above all, Cicurel, with its armies of overly deferential and overly educated salespeople, many of them Jewish, and floor after floor of French and Italian fashions—silk blouses, designer hats, leather bags, bolts of imported fabrics.

My first winter coat from Cicurel with the lone gray button and matching gray wool scarf was the loveliest I had ever owned. I couldn't bear to throw it away, even years later, when I had outgrown it and it was too tight and too short, so that my mother in her infinite compassion finally took it from me, folded it neatly, and placed it at the bottom of one of the twenty-six suitcases for storage.

"Un de ces jours," Edith would sigh; One of these days. It was her favorite saying, and it applied as much to the day we would retrieve her wedding gown as to when we would dig out my Cicurel coat as to when we would at last be able to return to Egypt.

Cicurel, Benzion, Hannaux—gone except for the buildings they'd occupied, shadows of their former selves.

They were like Cairo itself, haunted remnants of a city, both alive and dead.

My driver continued the sentimental journey through downtown Cairo, and then I spotted it: Groppi's—part patisserie and part paradise. I ran out of the car. It was at that moment, as I headed toward Groppi's door, that the feeling from the plane returned, the sense that Dad was at my side. As I walked in, I sensed his halting footsteps and instinctively slowed down, aware he had trouble keeping up.

It seemed, at first glance, exactly as we had left it. The stately structure with the delightful sign, "Groppi's" in fanciful longhand, a child's scrawl, still dominated Suleiman Pasha Square. Inside, the large room with tall ceilings and stately columns, pink walls and countless *étagères,* had once promised a palace of childhood—and adult—pleasures.

There were no customers inside. The shelves, once laden with distinctive pastries, were nearly all barren. The area in the front that once housed a thriving take-out business had a forlorn, abandoned look to it. Someone was manning the old wooden cashier station, but there was nobody in line. Like Cairo—like my family after Cairo—the famed establishment was all about decline and faded splendor.

I could almost feel my father frowning at the few trays of gaudy, thoroughly unappetizing pastries.

Where were the famed buttery desserts so light and delicate they could rival those of a Parisian bakery? And what of the crowds who would line up to purchase them, or sit in the café—elegant Italian women and their British officer-lovers, or all the others, the Greeks, the Belgians, the French, the Jews, in all their finery, who had made Groppi's the most cosmopolitan and decadent pastry shop in the world?

There were no menus anymore, and not much to order in any event. An old sign with an arrow pointing upward that read "Restaurant" now led to nowhere. The swank second-floor eatery where my father had rung in countless New Years dancing the tango and the fox-trot as a full orchestra played was bolted shut.

An Arab woman in a black chador that covered all but her eyes sat down at the table across from mine and ordered an espresso. I wondered how she would possibly be able to sip her coffee with the heavy black veil over her face. "Once upon a time, Arabs weren't allowed into Groppi's," my chauffeur told me, when I returned to the car. "Only colonialists went inside."

He sounded vaguely angry and reproachful, and I shivered to think I had been a six-year-old colonialist usurper. The revolution had taken care of all of that. Anyone could walk into Groppi's now. With the foreigners gone, every Egyptian could have their morning coffee there if they wished.

Few did. It was now a museum to a bygone era.

My driver, who had made me feel so guilty moments earlier, noted my dismay. He told me soothingly that I'd surely like the *other* Groppi's much more, the one with the garden. We could go visit it another day, he offered. That was another quality unique to Cairo: if despair was all around you, even so, hope was around the corner.

Despite its ruined state, Cairo was perpetually optimistic. It was like a genie's lamp, and if you only rubbed hard enough and long enough, then—voilà—it would deliver all that you had wanted these many years, the house where you'd grown up, the synagogues where you'd prayed, the stores where your parents had shopped, even the flowers whose heady scent had followed you across oceans and time all the way to New York.

ONCE IN MY HOTEL, I found that I couldn't sit still.

I rushed to the lobby and asked if I could hire a driver. I was intro-duced to Ahmed, a kindly Egyptian driver fluent in English. I repeated my request.

He seemed to instantly understand. With my husband at my side, and my father, I prayed, hovering somewhere nearby, we entered his taxi for what turned into a short, twenty-minute ride.

Suddenly, I was back, back on Malaka Nazli.

I knocked on the large wooden door at number 280, and immedi-ately a man answered. Incredibly, he didn't seem at all surprised to see a complete stranger show up after all these years demanding access. He was an engineer, thoughtful and patient, and he greeted me as if I were a long-lost friend or relative. Welcome, he said very kindly. "My house is your house."

The marble tile floor where I'd sat with Pouspous and cried the day we left Egypt was the same, as was the large living room around which the house was structured, so that communication was flowing and con-stant and loneliness was unknown.

I had left an Egyptian and returned as an American. With my Amer-ican obsession with privacy—privacy over hospitality, privacy over love, privacy over friendship, privacy over familial bonds—I found myself frowning, puzzled as to how my mother had given birth to five children in a house where it wasn't possible to shed a single tear alone.

To my Western sensibility, Malaka Nazli was much too open. Was it possible to talk, work, study, make love, without everyone knowing your business? What was it like when Baby Alexandra died? Had there been even a small corner where my mom could mourn her in peace? Where my father could pray for her soul as it drifted out of Malaka Nazli?

As I sat with the man, who, like me, was born in this house, his mother arrived. A soft-spoken woman in her sixties, with her gray hair in a neat bun, she had been a young bride when her uncle and father-in-law had negotiated the purchase of the apartment from my father in the spring of 1963, only weeks before we left.

How thrilled she was to be moving to such an elegant building, lo-cated on a grand, lively boulevard. It made her feel hopeful about her new life, and the young man she had married, and the children she hoped to bear in this spacious, airy apartment.

She remembered the day she arrived. There was nothing left in the house. It was devoid of all furniture and decorations and appliances, with the exception of two items—a black telephone and a white bed. Her first act was to get rid of the metal hospital bed, my father's during the months of convalescence that had turned into years.

She vividly recalled its owner: though she had met my father only once, when her in-laws concluded the deal, the tall, handsome older man had made a strong impression. His hands shook as he signed the agreement, severing any claim to the apartment he had occupied for thirty years, where he had seen his mother die and his children born.

She could still see that—the tremor in Dad's hand as he gave away Malaka Nazli.

What she and her husband kept was the black dial-up phone. That was a true luxury in 1960s Egypt, when only the wealthy—or the well-connected—could reach enough important people and spread adequate amounts of baksheesh to inveigle a phone line in their homes. Together, the young couple hatched a plot. Devout Coptic Christians, with crosses all over the house, they prayed to God to forgive them for the small deception they were about to perpetrate.

Each month, posing as my father, the husband reported to the Ministry of Communication to pay the charges. He even took care of some back bills we had incurred in the month or two before we'd left. He was polite and pleasant, careful to pay on time and to the penny, so that the disguise worked for years, until an astute bureaucrat realized what had happened, and took ruthless and immediate measures. He dispatched his workers to the ground-floor apartment to yank away the black phone and interrupt the service.

It would be years before the house had a phone again.

The Old Bride rose slowly from her chair, went to a drawer, and retrieved several pieces of paper covered with Arabic writing. They were invoices as well as letters from the Communications Ministry addressed to my father about "his" phone. Somehow, it didn't surprise me that she had kept them, as if anticipating the day one of us would come to settle all outstanding accounts—in the same way that she and her son hadn't seemed particularly surprised to see me at their doorstep.

Although the apartment was dilapidated and neglected, almost as if its occupants didn't care anymore, she graciously offered me a tour. Our first stop was the dining room, with the little balcony overlooking Malaka Nazli where I had spent so much of my childhood. There was a cactus plant to ward against the evil eye, and two small chairs placed on either side of the balcony; I was sure they had been ours, but she smiled, and shook her head no, and repeated, so that I had to believe her: only the phone and the bed had been left behind, nothing else. She loved to sit in the sun with her husband, and she had bought the chairs herself. Since her husband's death a year or so earlier, her son had begun to join her out on the balcony, and they'd continued to relish the street life.

It was still possible to do that, I realized to my amazement, even though decades had passed and Malaka Nazli itself had changed. Years back, Egyptian bureaucrats had decided to build a highway parallel to it, ostensibly to improve traffic flow. Now, the ugly concrete struc-ture extended like a dark shadow across the once-serene avenue, traffic

Loulou on the balcony at Malaka Nazli, next to the protective cactus, Cairo, 2005.

flowed only one way instead of two, and bottlenecks were worse than ever.

In my father's old room, the window where he and I had spent hours laughing and calling out to pretty girls and friendly passersby, chatting with anyone who would chat with us, was all boarded up; it was now the son's room, and he preferred to keep the window closed.

I shuddered to think what my father would have done—no doubt, he would have headed straight to the window and flung it open and let everyone, from the new owners to the Egyptian government, know that it was his street and his window, and he was reclaiming both.

We continued our tour with a visit to the kitchen with its modern gas stove and the refrigerator positioned a few feet away. Where was the Primus? I asked. What had they done with Grandmother Zarifa's beloved Primus?

The Old Bride seemed puzzled. Then she burst out laughing. Even before moving in, she had insisted on a brand-new kitchen. Out went the old-fashioned contraptions that had remained unchanged since the 1940s, when Zarifa reigned—the gas and kerosene burners, the icebox that masqueraded as a refrigerator, the corroded wooden shelves, and, yes, the Primus. Instead, the kitchen was outfitted with a real stove, electric, new countertops and cabinets.

I looked at my beaming hostess, and praised her sparkling and modern forty-year-old appliances.

And the bathroom, I wanted to know: Still no hot running water?

It was the son's turn to chuckle. For years after we'd left, his family didn't have hot water either. The two of us shared memories of childhood baths using an oblong metal container filled with boiling water our mothers would lug into the small bathroom. We'd fill a large mug with the water and splash it all over our head and body, simulating a hot shower.

"I'd cry out to my mother to heat more water because I kept running out," the Engineer recalled, and with those words, I suddenly remembered my ritual Friday-night bath on Malaka Nazli and how safe and protected I felt in that cozy bathroom, with the steam rising from the aluminum container, and my mother scrubbing and scrubbing my hair with Savon Nabolsi, the large green medicinal soap, because shampoo was too luxurious an item to be squandered on a little girl, and how delighted I was when she tossed cup after cup of hot water over my head and back.

Some years ago, the Engineer told me, the landlord had installed hot running water on demand, and it was possible to take real showers now. But he didn't look nearly as excited as he had moments earlier, when he'd recalled his little tantrums, his demands that his mother deliver continuous amounts of steamy water.

The master bedroom was tucked away in the back. It was the room my mother had briefly shared with my father after they wed, until he'd moved out and returned to his old digs, the airy room in the front facing Malaka Nazli, and his old ways.

Despite its many windows, the bedroom struck me as dark and dreary. It was where my mom had given birth to all of us, including Baby Alexandra. It was the only part of the house, the only part of Cairo, where I didn't want to linger, which felt impossibly bleak.

At last, we came to my favorite room, the one overlooking the alleyway. It had once been Zarifa's room, and years later, my father's office. I'd loved to play there, amid the papers and files, or better yet, to stand on the balcony with Pouspous at my side, waving and chatting with my friend across the alley, the pretty bride.

It was also where the vendors would station themselves each morning, balancing baskets of fresh fish on their heads or pushing carts filled with the fruits and vegetables du jour—zucchini, cucumbers, green beans, potatoes, apricots, or my favorite, the dark purple baby eggplant. The balcony was so close to the ground that we could reach out and touch the baskets on their heads. I could point to what I wanted, help myself to bunches of grape leaves that my mother would stuff with ground meat and rice, and simmer in a lemony broth.

Who needed to go to the supermarket, anyway? Once upon a time, in Cairo, the supermarket had come to us.

I had loved the call of the vendors as they approached Malaka Nazli—shrill, intense, designed to make sure everyone heard them coming. Their high-pitched song had followed me all the way to America, much as the scent of the roses had pursued my father.

My mother had tried to pull me away from the balcony. She'd worry when there were funeral tents, and parades of mourners in the alley below: she didn't want me to know death or sadness, no doubt trying to stave off the day when I would know both only too well. Yet I had rarely felt sad there, even when the mourners wept out loud. Only when my friend the young bride died, and I didn't understand.

The Old Bride seemed troubled when I asked her about the apartment across the alleyway. When she'd first arrived, there was a man, a solitary figure who lived there. He was a military officer, often in his uniform, who occasionally made an appearance on the balcony. But he never smiled, or acknowledged her greetings with more than a curt and somber nod. He was a widower, she thought, and lonely, but no matter how she tried, he wouldn't respond to her attempts at friendship. Perhaps he was only being proper, she decided. She was, after all, a newly married woman.

Then one day he moved out, suddenly and without any notice. The bad-luck apartment remained empty for years. Finally, after no one had moved in, an art-store owner took it over and made it part of his gleaming new gallery.

"Do you remember a cat, a small cat of many colors, very sweet and affectionate, who would have been here when you moved in?" I asked.

The Old Bride turned to her son, thoroughly bewildered. "Otah? Otah?" she said, repeating the Arabic word for cat to make sure she understood. She was such a kindly woman, deeply instinctive, and blessed with a good heart: Who else would have saved some forty-year-old bills for a black phone? It seemed entirely plausible to me that she would have taken Pouspous in, fed her cheese and sardines, exactly as my father had told me when we left.

But the woman kept asking, "Otah, otah?" She was incredulous; in her mind, the conversation had taken a very strange turn.

Here someone had come all the way from America, and she was inquiring about some long-forgotten cat.

The Old Bride shook her head decisively: No. There was no cat in the house when she'd moved in, a few weeks after we'd left Egypt. There had been only the white bed and the black phone.

I wandered over to my father's old room. I suddenly felt like crying. I thought of all the stories I'd carried in my head these many years, stories he had told me after we had left, when I was feeling so forlorn—about Pouspous doing well, enjoying herself in the house we'd left behind, sunning herself on the balcony and inviting strangers to feed her. None of it had been true.

Both mother and son could tell I was distraught. I sat down as she hurried to serve me a cool drink from her modern refrigerator. Then she excused herself and left.

"You know, of course, what happens when a cat must lose its owner?" the Engineer asked me cryptically, in his labored English.

I shook my head, no.

"When a cat can no longer find its owner, it stops eating—it stops eating completely," he explained. "It is as if they are in mourning. That is how they die," he said gently. "They die after refusing to touch a single morsel."

THERE WAS A NEIGHBOR upstairs that I should meet, the Old Bride told me when I returned the following day to Malaka Nazli.

She had lived in the building almost sixty years and had known everyone who had come and gone, including my parents; now, she was very anxious to meet me, but she was too aged and infirm to walk down even the lone flight of stairs. Would I mind going to the second floor to see her?

Reluctantly I left the apartment.

My apartment.

As I marched up the stairs, I noticed how dusty and broken down and neglected they were, the walls blackened and filthy, the floors looking as if they hadn't been swept in years—perhaps in all the years since my family had left.

I'd played with Pouspous on this stairwell when it sparkled, chasing after her as she ran to hide in the thousand nooks and corners only she knew. I'd often had no choice but to enlist the aid of Abdo, our Sudanese porter, to find her. Abdo lived downstairs, in a dark, mysterious basement apartment directly below ours. He seemed always there when we needed him to hail a taxi for my father or run an errand for my mother, or help me hunt for Pouspous because he knew the secrets of the stairwell.

"Abdo, Abdo," we'd call out, and he would materialize out of the darkness, smiling and gracious in his flowing white caftan, strangely dignified.

Abdo was long gone and had never been replaced. Now, like so much of Cairo, the building had been allowed to decay until little by little, it became dirty and unkempt and lost much of the elegance and grace that had prompted my father to move in with Zarifa and Salamone in the spring of 1938, and then bring his new bride, my mother, to live there five years later.

I knocked on the door and was greeted by a young woman in traditional Arab garb. She welcomed me into the apartment, which was a simulacrum of our place downstairs—the same open design, the same four bedrooms built around the central room, and off to the side, the narrow kitchen.

Her mother sat in a velvet armchair, a frail, regal figure with her hair swept under a white head covering. I went over to shake the old woman's hand, but she quickly reached out to embrace me instead, her arms wrapping around me as she kissed both my cheeks, and brought me close to her chest. She looked as if she could barely stand without an effort, yet her gaze was focused and strong and not at all vague in the manner of the old.

I could feel her eyeing me closely, studying my eyes, my face and hair, examining my clothes and my shoes, as if trying to remember, to remember.

I sat on the sofa directly across from her. She kept looking at me, not saying a word, while her daughter kept up a light banter. Would I like a cold drink? A bit of dinner? There was some nice okra stew cooking in the oven: could she offer me a plate? Was I enjoying my visit to Cairo?

Suddenly, the old woman interrupted us and began to speak.

"You look exactly like your mother," she declared in Arabic, for she knew no other language. "You are the same as her."

She paused to take a sip from a glass of tea. "She was so little—very little—and your father was big, much too big."

She recalled my mother as soft-spoken and delicate, and above all as someone who loved children. "She would always give candy and chocolates to my girls. Remember?" she said, turning to her daughter, who smiled and nodded, though it was by no means evident that her memory was as clear as the Old Bride's.

Suddenly, I could see my mother, Edith, as a young housewife, fishing in her purse for some bonbons, because she was good-hearted and liked spoiling children and missed her days as a teacher at the École Cattaoui where she had been the popular Mademoiselle Edith, object of respect and adulation.

I walked over to the old woman and took her hand to kiss it.

"Do I really look like my mom? Are you absolutely certain?" I asked her.

I was looking in the large mirror in the center of the living room, near her, studying my face and features and praying that her answer would be yes.

I felt her staring at me intently all over again.

"Absolutely—except for the teeth," she said, frowning. "The mouth." I could tell that she was sifting through the labyrinth of her mind with its eighty years of stored-up memories and impressions, trying to discern what was different about the woman in front her and the woman she'd met some fifty years earlier.

She was smiling now because it had all come back to her, it had all come back perfectly.

We continued to chat; my driver was translating, though I no longer needed him: I felt that I understood her perfectly by her gestures and smiles. Her daughter gave me the obligatory tour of the house—this bedroom, that sitting room; they were all deserted. Finally, she beckoned me over to the balcony, which was the family's pride and joy, with its graceful concrete canopy, its panoramic views of Malaka Nazli and beyond it, Cairo itself.

I put my jacket on, smoothed my hair, and rose to leave. The old woman suddenly cried: "Wait."

I stopped to look at her.

"I am old and I am lonely," she cried. "There is only me and my daughter here, and I have so many rooms." With a sweeping gesture, she pointed to the empty rooms, the dining room devoid of any diners, the bedroom without a husband, the sitting rooms and playrooms with no children.

"Why don't you stay?" she said. "Why don't you move here?"

I looked at her, after my driver had translated what she'd said, and translated it yet again.

"You can have any room you want," she added, sensing my confusion, though not really understanding it. To her, it was perfectly natural to ask this stranger, who really wasn't a stranger at all, who was as familiar to her as her own past and her own family, to come live with her.

I was being offered a chance to move back: to move back to Malaka Nazli.

I ran to embrace the old woman. That moment, when she held my hands in hers, I suddenly understood my father, and his despair after Cairo, and the sense of desolation that he had tried to blame on the flowers that didn't have a scent and the people who didn't have a heart.

Malaka Nazli hadn't simply been a place, I realized, but a state of mind. It was where you could find an extraordinary, breathtaking level of humanity. What it lacked in privacy, what it failed to provide by way of modern comforts—hot running water, showers, electric stoves, refrigerators, telephones—it more than made up for in mercy and compassion and tenderness and grace, those ethereal qualities that make and keep us human.

If Adly Street had been the way to the Gates of Heaven, then Malaka Nazli was paradise itself, and Dad had been fortunate enough to taste it, and I was lucky to glimpse for myself what he'd meant all these years when he kept his small suitcase packed and ready to go.

As I climbed back into our car, I glanced up to see the old woman standing on her balcony. She seemed forlorn and lost in thought, looking out to the farthest reaches of Malaka Nazli. She was searching up and down the boulevard, as if trying to find me, as if trying to find not only me but also my parents, and her husband, and her own youth— that time when she was a girl on a balcony with a family waiting for her to come inside.

As we drove away, I felt that I was leaving all I cared about behind, not simply a stranger who had shown me such unexpected kindness, but another old woman, my grandmother Zarifa, and another, Nonna Alexandra, and a young woman, too, my mother, Edith, crossing the threshold of Malaka Nazli as a twenty-year-old bride, and Baby Alexandra, the sister I had never known, and my two uncles who had seemed forever lost—the child of the souk and the priest, returned from his Jerusalem monastery—and my aunt Bahia, back from Auschwitz, clutching her husband and Violetta, and my father, above all, my father.

I felt as if they were all standing there, on that wrought-iron balcony.

ACKNOWLEDGMENTS

I n the spring of 2004 I found myself leafing through my family's files obtained from HIAS, the Hebrew Immigrant Aid Society.

Amid the yellowing papers, I stumbled on a ledger that chronicled my
father's repayment of the debt we incurred coming to America on the *Queen
Mary*—a month-by-month accounting of each of my dad's ten- and fifteen-
dollar checks.

I proposed a piece about my father and the ledger to Erich Eichman, the
Wall Street Journal books editor who oversees cultural commentary, and to my
great delight he decided to publish the column on Father's Day 2004.

He was a deeply sensitive editor, and I owe him so much.

The day the piece appeared, I was flooded with calls and e-mails. One of
the callers was literary agent Tracy Brown who felt there was a book to be
done about Leon and my relationship with him. From the start he was insightful and supportive—helping me to frame the story and offering me the constant feedback that carried me through this undertaking. *The Man in the White
Sharkskin Suit* would not have been possible without Tracy's luminous com-
ments, grace, and sensibility.

My encounters with Dan Halpern, Ecco's publisher, showed me why he is
such a legendary figure and my book found such a lovely home. I was moved

by his passion for the Levant—from the music of Om Kalsoum to the rose water in Oriental pastry.

At Ecco, I am indebted to acquiring editor Julia Serebrinsky for her enthusiasm: As a Russian immigrant, she seemed to relate completely to the story of my Levantine father.

Lee Boudreaux, Ecco's editorial director, who oversaw every aspect of the book, proved to be an extraordinary editor and friend: Tireless, charming, rigorous, weekend after weekend she would take my manuscript home and figure out precisely what needed to be done. I am profoundly grateful both for Lee's passionate dedication and sense of excitement.

Ghena Glijansky was an invaluable assistant editor who gave careful, loving guidance and helped fine-tune the book even after she left Ecco.

Editorial assistant Abigail Holstein taught me the meaning of the word "indefatigable." Art director Allison Saltzman created the most exquisite cover.

A number of relatives generously shared their knowledge of my tangled family history over the book's hundred-year span.

One of the greatest joys of working on this memoir was being reunited with Salomone Silvera, my cousin in Milan who had lived with my father in Cairo from 1937 to 1944, and was able to provide exquisite details about life inside Malaka Nazli. In meetings at his home near the Duomo, Salomone conjured up lost family members, including my Syrian grandmother, Zarifa, who through his colorful stories became a major character. He taught me much about his own mother, my aunt Bahia, who had perished at Auschwitz together with his father and sister. I will always be indebted to Salomone and his wife, Sally.

My oldest brother, César, emerged as the keeper of the flame, the family archivist who collected my father's papers for more than forty years—business cards, Leon's wallet, even old canceled checks from some of his favorite charities. César also volunteered his own memories of Egypt, France, and our earliest days in America, hilarious and sad at the same time. He unearthed most of the photos in this book. I am grateful to him and his wife, Monica.

My sister, Suzette, vividly described Alexandra, our beautiful, gifted, and supremely sad maternal grandmother. She even sang for me the Italian songs Alexandra had loved, and provided useful insights into the world of Egypt's Jewry after Suez.

David Ades, my cousin in Los Angeles, offered charming recollections of how his mother, Tante Rebekah, prepared rose petal jam. Victor "Pico" Hakim regaled me with stories of my father's work in Cairo and volunteered details about his mother, Tante Rosée, and other members of my mother's family. He and his wife, Rachel, were deeply kind and hospitable.

Josette Hakim gave heartbreaking descriptions of Alexandra lost amid the orange groves of Ganeh Tikvah.

I also wish to acknowledge the wondrous Desi Sakkal, founder of HSJE—the Historical Society of the Jews of Egypt, the website that has enabled Egypt's lost Jewry to begin to reclaim their heritage and to finally reconnect. Desi was a constant resource, deeply caring, always eager to answer my most obscure questions. His organization is nothing short of miraculous. Albert Gamill and Dr. David Marzouk, fellow expatriates, were deeply kind to read over my manuscript and review every one of my Arabic phrases.

Two authors of works on colonial Egypt must be singled out for special praise. Samir Raafat's lively, witty books and articles were invaluable to me, and Artemis Cooper's wonderfully detailed and charming book on World War II Cairo was a terrific resource.

I am grateful to the Egyptian government for welcoming me back to Cairo and offering me access to Jewish sites. Aaron Kiviat, a young American who lived and studied in Egypt, helped me plan my return. Carmen Weinstein, who carefully oversees what's left of the Cairene Jewish community, kindly opened the doors to the Gates of Heaven and allowed me to visit the shrine of Maimonides.

Arnold Paster of Southampton, New York, was a grand and magical friend, encouraging me in the most difficult stages, convincing me that I could really accomplish this, and even surprising me with a bottle of Dom Pérignon when I completed the first draft. The essence of style, with a special fondness for wearing elegant white suits, Arnie became a kind of muse. He also helped me come to terms with my enigmatic father and to embrace his extremes and contradictions.

Rabbi Rafe Konikov and his wife, Chany, offered me a home away from home at the Chabad of Southampton Jewish Center. It was a haven, a place where I could reflect and where, little by little, ideas jelled: In the course of an ordinary Saturday-morning service, I would find myself thinking of Leon, and my thoughts were like a prayer. Once I looked up at the synagogue's bay windows and swore I caught a glimpse of him in the garden, wearing one of his jaunty hats and looking my way. Afterward, I would strain to see him again, but I never did, except on the pages of my book, where I saw him constantly.

At the *Wall Street Journal,* no one was more encouraging or supportive than Paul E. Steiger, the managing editor who hired me. He has consistently been both mentor and friend. I revere his instinct and his intellect and all-abiding love. The *Journal*'s Daniel Hertzberg was deeply generous in granting me leave to work on the book, as was my editor, Joe White.

Arthur Gelb, the grand old man of the *New York Times* and its former managing editor and cultural czar, was wonderful and loving as only he can be and helped keep me focused—I cherish our friendship and deep bond.

Ken Wells, my former editor at the *Journal* and the author of many books himself, was profoundly encouraging to me throughout every step; he is the most exquisite and gifted role model and friend I can ever hope to have.

Doctors are a crucial part of this book and of my world. I am, as ever, grateful to Dr. Burton Lee, who was able to re-create scenes and events that took place more than thirty years ago. I consider Dr. Lee one of the great men on this earth, and I love him utterly and completely. I am also indebted to Dr. Ronald Schwartz who helped me bring *Sharkskin* to fruition. Finally, I owe much to my internist Dr. Jerome Breslaw, who has tried so hard to make me and keep me well.

Steve Solarz had wonderful insights into the Syrian-Jewish community he had come to know as their congressman.

Steve Olderman, a creative director at R/GA, was deeply generous in sharing ideas about the jacket and marketing of *Sharkskin*.

I want to convey special gratitude to the sublime Grace Edwards, the novelist of lost Harlem, and my teacher at Marymount's Writing Center. It was Grace who heard the earliest iterations of *The Man in the White Sharkskin Suit* and taught me the value of reading one's prose out loud. She is kindness incarnate. I am indebted to Lewis Frumkes, the Center's director, for creating such a nurturing place. I am also grateful to my wonderful class at Marymount, who heard me read chapter after chapter and offered so much insight and help.

Clifford David was such a magnificent friend. He is a wonderful actor and a gentle force of nature.

Maryann Callendrille and Kathryn Szoka turned Canio Books in Sag Harbor into my favorite literary retreat on earth. In Southampton, Jack Biderman was deeply kind and encouraging; the most wonderful friend imaginable. My childhood friend Stella Ragusa was generous with her memories of her family and mine. George Getz and Sandy Becker, my New York accountants, gave me terrific financial counsel. Silvia Burgos and Kyle Spelman at the *Journal* gave me the technical support I needed to work. Daniel Pipes graciously answered my thousand questions about Cairo.

When I returned from Cairo in the spring of 2005, I visited Rabbi Raphael Benchimol of the Manhattan Sephardic Congregation and told him of the strange sensation that overcame me from practically the moment I was airborne, and continued as I explored the streets of my childhood—the feeling that my father was by my side.

A mystic at heart, the rabbi didn't seem surprised. "We are taught that although people leave a place, their effect remains imprinted and permeated there for eternity. Their footsteps, going from their home to the synagogue and back, remain imprinted forever. Indeed," he said, "even the air they occupied and the contour it shaped remain forever."

I would never have been able to work on *Sharkskin* without my husband,

Douglas Feiden, who listened to me read out loud every chapter—every word—that I had written, and offered profound, discerning, stern but always loving guidance. He accompanied me to Paris, Milan, Brooklyn, and Cairo and was with me when I knocked on the door of 281-Malaka Nazli Street, sharing my joy and my pain at being home finally and again.

SELECTED BIBLIOGRAPHY

BOOKS

Aldridge, James. *Cairo: Biography of a City*. Boston: Little, Brown, 1969.

Beattie, Andrew. *Cairo: A Cultural History*. New York: Oxford University Press, 2005.

Benin, Joel. *The Dispersion of Egypt's Jewry*. Berkeley and Los Angeles: University of California Press, 2004.

Cooper, Artemis. *Cairo in the War, 1939–1945*. London: Hamilton, 1989.

Danielson, Virginia. *"The Voice of Egypt": Umm Kulthum, Arabic Song, and Egyptian Society in the Twentieth Century*. Chicago: University of Chicago Press, 1999.

Heikal, Mohamed. *The Cairo Documents: The Inside Story of Nasser and His Relationship with World Leaders, Rebels, and Statesmen*. New York: Doubleday, 1973.

Hopkins, Harry. *Egypt: The Crucible—The Unfinished Revolution in the Arab World*. Boston: Houghton Mifflin, 1970.

Raafat, Samir W. *Cairo, the Glory Years: Who Built What, When, Why and For Whom*. Alexandria, Egypt: Harpocrates, 2003.

Rodenbeck, Max. *Cairo: The City Victorious.* New York: Alfred Knopf, 1999.
Stadiem, William. *Too Rich: The High Life and Tragic Death of King Farouk.* New York: Carroll & Graf, 1991.
Wilber, Donald N., ed. *United Arab Republic, Egypt: Its People, Its Society, Its Culture.* New Haven, Conn.: HRAF Press, 1969.

NEWSPAPERS AND MAGAZINES

Al-Malky, Rania. "Where the Streets Have No Name." *Egypt Today,* April 2005.
Eban, Suzy. "A Cairo Girlhood." *New Yorker,* July 15, 1974.
Hassan, Fayza. "In the Pashas' Den." *Al-Ahram Weekly Online,* no. 459 (December 9–15, 1999).
———. "Sent Away: Who Was King Farouk?" *Al-Ahram Weekly Online,* no. 572 (February 7–13, 2002).
Heard, Linda. "Groppi: People's Memories of the World's Ritziest Tea-Room." *Community Times,* October 2006.
Raafat, Samir. "Gates of Heaven." *Cairo Times,* September 2, 1999.
———. "Groppi of Cairo." *Cairo Times,* June 15, 1996.
———. "Resurrecting Street Names." *Cairo Times,* May 11, 2000.
Sanua, Victor. "The Vanishing World of Egyptian-Jewry." *Judaism,* Spring 1994.
Shaimaz, Fayed. "Downtown Cairo: Egypt's Bohemian Rhapsody." *Community Times,* November 2006.

WEBSITES AND OTHER NEW MEDIA

Bassatine News. Online Jewish newsletter from remaining Jewish community in Cairo. www.geocities.com/rainforest/vines/5855/bassai.htm.
"The Golden Age of Egyptian Dance," "Taheya Carioca aka Tahiyya Karioka," and "Samia Gamaal." Belly Dance Museum. *www.belly-dance.org/* and *www.venusbellydance.com.*
IAJE, International Association of Jews from Egypt. *www.iaje.org.*
Kiviat, Aaron. "I Buried My Father's Talis at Bassatine." Letter dated November 25, 1999. Historical Society of Jews from Egypt. HSJE.org (posted 2006).
Sakkal, Desire, ed. "General News and Information." Historical Society of Jews from Egypt. *www.HSJE.org.*

MANUSCRIPTS AND ARCHIVES

Lagnado family file. HIAS (Hebrew Immigrant Aid Society), Paris and New York City, 1963–79. Courtesy HIAS.

Lagnado family file. NYANA (New York Association for New Americans), New York City, 1964–65. Courtesy NYANA.

Lagnado family file. COJASOR social service agency. Courtesy of Mémorial de la Shoah—Musée, Centre de Documentation Contemporaine, Paris.

Curriculum Vitae, Death Certificate of Salomon Lagnado aka Père Jean-Marie Lagnado. Courtesy Davide Silvera, Israel.

Red Cross. Letters to Salomone Silvera on the deportation of Bahia, Lelio, and Violetta Silvera to Auschwitz. Cairo, 1945–47.